52 Topics in Current Chemistry

Fortschritte der chemischen Forschung

W0036563

Medicinal Chemistry

Springer-Verlag
Berlin Heidelberg GmbH 1974

This series presents critical reviews of the present position and future trends in modern chemical research. It is addressed to all research and industrial chemists who wish to keep abreast of advances in their subject.

As a rule, contributions are specially commissioned. The editors and publishers will, however, always be pleased to receive suggestions and supplementary information. Papers are accepted for "Topics in Current Chemistry" in either German or English.

Any volume of the series may be purchased separately.

Library of Congress Cataloging in Publication Data
Main entry under title:

Medicinal chemistry.

(Topics in current chemistry 52)
1. Chemistry, Medical and pharmaceutical-Addresses, essays, lectures. I. Series.
[DNLM: 1. Chemistry, Pharmaceutical.   W1 T0539M v. 52 / QV744 M489]
                            RS402.M42   615'.19   74-12422

ISBN 978-3-662-15548-6   ISBN 978-3-540-37834-1 (eBook)
DOI 10.1007/978-3-540-37834-1

© Springer-Verlag Berlin Heidelberg 1974
Originally published by Springer-Verlag Berlin Heidelberg New York in 1974
Softcover reprint of the hardcover 1st edition 1974

Contents

Design of Bioactive Compounds

Prof. Dr. Everardus J. Ariëns and Anna-Maria Simonis

Pharmacological Institute, University of Nijmegen, Nijmegen, The Netherlands

Contents

1

1. Introduction

This paper deals with the design of bioactive compounds including drugs, food additives, pesticides, industrial chemicals, etc., in general, all foreign compounds to which biological objects — man, animals, and plants — may be exposed. Nutrients are excluded unless used in nonphysiological dosages or administered by a nonphysiological route. The word "pharmacon" (Greek $\varphi\alpha\rho\mu\alpha\kappa o\nu$, meaning "active principle") is used to indicate these bioactive compounds.

The exposure of living organisms to pharmaca involves both desirable and undesirable effects. The term "internal pollution" is used to describe interactions between pharmaca and living systems that operate to the disadvantage of man. Internal pollution is generally a result of external pollution, including improper use of drugs, food additives, etc. The determinant is the balance between advantage and disadvantage. The primary goals in the design of new pharmaca are therefore optimization of desired effects and minimization of undesired effects.

The word "design" stands for the efforts to reach these goals on as rational a basis as possible, which implies that the trial-and-error factor must be reduced to the feasible minimum. It involves modulation of the biological actions of particular agents by molecular manipulation[12]. This requires a certain insight into the relationship between structure and action, or better still, into the relationship between the physicochemical properties of the agents concerned and their biological actions, both desired and undesired. The discipline involved in the design of pharmaca on this basis can be called "pharmaco-chemistry".

1.1. Bioactivity with the Emphasis on Toxic Actions

The results of exposing biological objects to pharmaca can include:
A) Quick passage of the compound through the biological object without affecting its function.
B) Accumulation, possibly sequestration, of the compound in the biological object without affecting its function.
C) Induction in the biological object of reversible and therefore temporary effects, which disappear with the elimination of the compound. Most therapeutic effects of drugs and many of their side-effects belong in this category.
D) Induction of irreversible changes in the biological object by a reversible interaction with the pharmacon, such that the effect continues even after the drug is totally eliminated. An example is abnormal embryonic development due to early intrauterine exposure to hormones.
E) Induction of irreversible changes in the structure of essential cell constituents as a result of exposure to the pharmacon. This implies a certain degree of chemical denaturation of these cell constituents indicated as "chemical lesions"[3, 4, 24].
Chemical lesions can be differentiated as follows:
a) chemical interactions between the pharmacon or its metabolic product(s)

3

and cell constituents such as DNA, RNA and proteins with covalent bond formation[5-13, 165, 168, 170, 172];

b) chemical changes induced in the cell constituents by peroxides formed during the process of pharmacon metabolism. This implies that chemical changes occur in the cell constituents but without forming chemical bonds with the pharmacon or its metabolites[14, 15, 169];

c) incorporation of the pharmacon or its metabolite(s) into the cell constituents as metabolic but afunctional analogs. This process, known as lethal synthesis[16, 98], is not uncommon in the action of antimetabolites.

Reversible interaction with the biological object results in pharmacodynamic effects which, in certain circumstances, may be undesired and therefore must be classed as toxic. Chemical lesions, with very few exceptions, are classed as definitely toxic.

The toxic effects due to chemical lesions include:

a) carcinogenesis, involving chemical lesions in DNA[5-13, 103, 168, 170];

b) mutagenesis, also involving chemical lesions in DNA[17-20, 81, 104, 167, 168, 170];

c) teratogenesis, caused by disturbed cell proliferation due to chemical lesions during embryogenesis[21-23, 82];

d) possibly accelerated aging, caused by an increase in the error frequency in nuclear DNA[25, 26];

e) allergic sensitization, producing chemical lesions in proteins that cause them to act as allergens[27-31, 161];

f) cell degeneration and necrosis, due to chemical damage to the membranes of lysosomes or to essential enzymes[20, 32-36, 44, 105, 166];

g) photosensitization, involving formation of reactive products by radiation of the pharmacon or its metabolite(s) causing local chemical lesions or the formation of allergens[30, 37-40, 106];

h) chemical lesions which, if restricted to parasites such as pathogenic microorganisms and cancer cells, may be advantageous to the host.

Regeneration processes, such as substitution of denatured enzymes by enzyme synthesis and mechanisms for the repair of DNA, can sometimes render the effect of the chemical lesion transient. On the other hand, taking into account the essentially irreversible character of chemical lesions, there is a tendency for the effect to be cumulative. This accumulation is one of the main factors in the long-term toxicity of cancerogenic and mutagenic agents.

1.2. Main Phases in Biological Action

Three important phases can be distinguished in the action of pharmaca:
a) the exposure phase, or pharmaceutical phase;
b) the pharmacokinetic phase;
c) the pharmacodynamic phase (see Fig. 1)[41-43, 50].

a) The exposure phase, or *pharmaceutical phase,* comprises, for instance, the processes involved in the disintegration of the form in which the agent is ad-

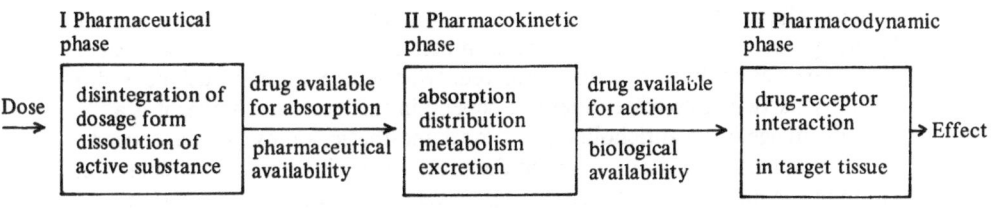

Fig. 1.

ministered and the dissolution of the active agent in the case of therapeutic tablets, spreading in the case of pesticides, retention in the case of exposure to air pollutants, etc. The fraction of the dose available for absorption is a measure of pharmaceutical availability or the efficacy of exposure. When taking the time course of the events into account, the term "pharmaceutical availability profile"[41-43] can be used. This profile depends on the formulation, the route of administration, and the dosage regimen.

b) The pharmacokinetic phase comprises the processes involved in the absorption, distribution, excretion and metabolic conversion of the pharmacon. The fraction of the dose that reaches the general circulation is a measure of the biological availability. The relationship of time to the plasma concentration is called the "biological availability profile" and that of time to the target-tissue concentration is the "physiological availability profile"[41-43]. Preparations of equal bioavailability may differ completely in bioavailability profile and hence in efficacy.

Due to possible differences, especially in pharmaceutical availability, the fact that drug preparations have chemical equivalence – equal contents of active agent – in no way guarantees biological equivalence[45-48, 66, 67]. This nonequivalence is manifested in differences in the biological availability profile.

c) The pharmacodynamic phase covers the processes involved in the interaction of the pharmacon and its molecular sites of action (receptors) with the consequent induction of a stimulus, as well as the sequence of biochemical and biophysical events thus initiated, which finally results in the effect observed. The specific characteristics of the pharmacodynamic phase for particular types of drugs thus provides a classification based on their mechanism of action. The time-effect relationship, however, depends mainly on the bioavailability profile and the pharmaceutical availability profile[41-43].

1.3. Structure and Action

The processes that underlie the various phases in the action are both chemical and physicochemical in nature. The relationship between chemical structure and the effect induced in the parmacodynamic phase of drug action is characteristic for the particular mechanism of action and thus for the type of effect and the type of pharmacon involved. This is not the case for the relationship

5

between chemical structure and the various processes involved in the pharmaco-kinetic phase. This relationship depends much more on the general chemical characteristics of the compounds involved. Whether a compound will be hydro-lyzed by particular esterases does not depend in the least on whether it is a curariform agent, an anticholinergic, a local anesthetic, an insecticide, a weed-killer, or a plasticizer, but chiefly on the presence or absence of a suitable ester group in the molecule. The relationship between structure and absorption, distribution, and excretion (as far as passive transport processes are involved) depends mainly on the overall physicochemical characteristics of the com-pounds in question, *e.g.* the partition coefficient.

The chemical properties and hence the chemical structure of a compound definitely determine its participation in the partial processes making up the various phases of action. The relationship of structure to action is therefore a fundamental characteristic of the action of pharmaca. The apparent absence of such a relationship can only be due to deficient methods of investigation and to the multiplicity and complexity of the processes involved. The "struc-ture-action" relationship will be found to emerge more clearly if it is studied with regard to particular part-processes such as those involved in pharmacon metabolism, in pharmacon distribution, and in pharmacodynamics. Such a study involves the use of simple, isolated test systems.

Efforts to detect the structure-action relationship usually emphasize the significance of particular moieties in the pharmacon molecule for particular aspects of its action[1, 2, 49-51]. Chemical modification of the structure of a bioactive compound in efforts to modulate its action in practice always involves changing particular moieties in the molecule[1, 2, 43, 49-51], for instance by substitution[52-56].

1.4. The Design Procedure

The main phases in the action of pharmaca, as outlined above, are reflected in the design procedure. It starts with the recognition of or search for compounds having a particular type of pharmacodynamic action that is of potential use. These compounds then serve as pointers to the development of agents capable of inducing the desired type of effect with a proper time-effect profile and without undesirable effects. This profile and hence the biological availability profile and in turn the pharmacokinetics should be taken into consideration at an early phase of the design procedure. This implies that, if possible, not only the chemical modifications (low dose, high response) aimed at optimizing the compound in the strictly pharmacodynamic sense, but also modifications aimed at inducing the required pharmacokinetics must be taken into account. Modula-tion of pharmacokinetics may perhaps contribute to the modulation of the pharmacodynamic action as such. For instance, chemical changes that prevent penetration of the compound through the blood-brain barrier will prevent or eliminate actions affecting the central nervous system. Finally, formulations must be designed to optimize the dose-effect profile.

The possibilities and limitations of the design of bioactive compounds will now be discussed against the background of the three main phases of action. Various relevant aspects of the modulation of the pharmacodynamics and pharmacokinetics are considered.

2. Pharmacodynamic Aspects of Design

2.1. Biochemistry as a Source of Leads

One of the rational approaches in the design procedure is to search for new leads in the field of biochemistry and pathochemistry or, in general, to study the fundamental processes of life. This approach has been rather productive in the field of hormones, including plant hormones (auxins) and neurotransmitter substances, and has led to the development of compounds serving as substitutes for or as antagonists of the biological agents. The development of pheromones (insect lures) and of insect hormones opens up interesting perspectives. The increasing insight into the biochemical processes involved in the propagation of genetic information and the regulation of enzyme synthesis offers further possibilities, for instance, for designing analogs of repressor and derepressor agents[1].

The tremendous number of metabolic inhibitors developed in the past few decades is impressive. Thousands of analogs of substrates and intermediate products of the various biochemical pathways have been developed and tested[57-65]. They include compounds that act as competitors or that mimic the parent compounds not only as substrates, but also in the sense of inhibiting or stimulating feedback. The metabolic inhibitors are most useful tools in biochemistry. This approach, however, has resulted in agents suitable for practical application in only a few cases. As examples we can cite the organic phosphates and carbamates used as pesticides, and certain acetylcholine-esterase inhibitors and monoamine-oxidase inhibitors used as drugs. Besides the term metabolic inhibitors, the term antimetabolites is often used in relation to the design of structural analogs of substrates for enzymes (metabolites) and also in relation to analogs of vitamins, hormones, neurotransmitter substances, etc. It is sensible to restrict the terms metabolites and antimetabolites to metabolic substitutes (parametabolites) and the corresponding metabolic inhibitors (antimetabolites) and to use specific terms for other substances, i.e. hormones, hormonoids, and antihormones; vitamins, paravitamins and antivitamins; adrenergic and cholinergic neurotransmitters, adrenergic and cholinergic agents, and adrenergic blocking and anticholinergic agents; etc. In general, functional analogs capable of substituting for the biological product may be called mimetics and compounds acting as their competitive antagonists lytics[68].

2.1.1. Metabolites and Antimetabolites

Often there is a clear chemical relationship between the metabolite and its antimetabolites. Substitutions close to the group involved in the enzymatic

conversion of the substrate or metabolite often result in antimetabolite properties. α-Alkyl or *ortho*-alkyl substitution in esters, amines, amino acids, etc.
usually results in compounds that are resistant to and act as inhibitors of the
enzymes concerned (Fig. 2). A special class of antimetabolites is the "active-
site directed" irreversibly blocking agents[69−72]. These are substrate analogs

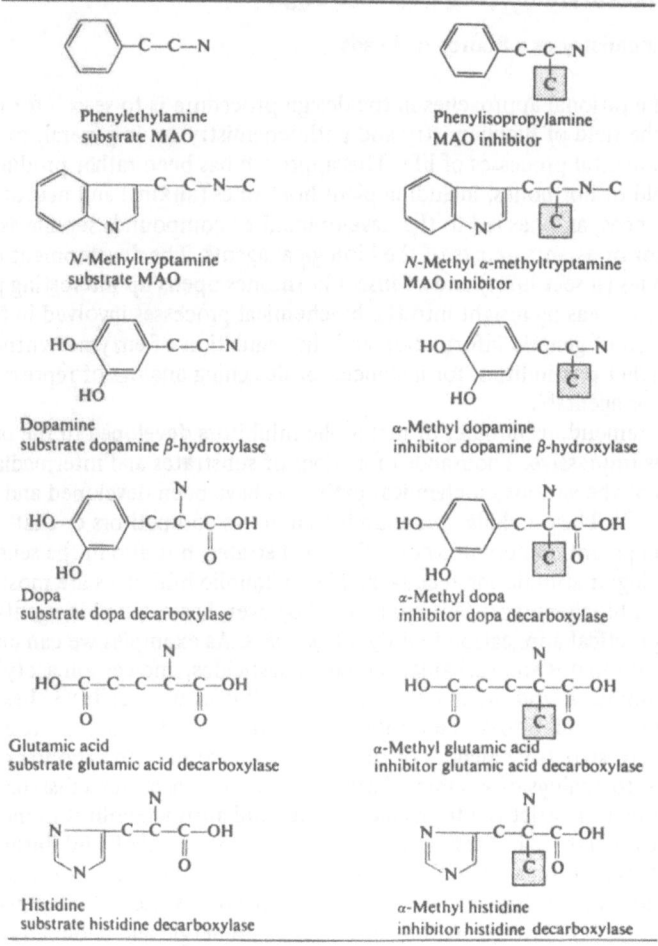

Fig. 2. Introduction of protecting groups resulting in compounds acting as enzyme
inhibitors

with alkylating or acylating moieties in a position such that they are capable
of forming irreversible covalent bonds with groups on the enzyme in the neigh-
borhood of the active site (exo-alkylation) or with groups constituting the ac-

tive site (endo-alkylation)[69]. The substrate molecule in this case serves as a moiety directing the alkylating groups to the target enzyme.

Why then, since such an abundance of metabolic inhibitors is available, do so few of them find practical application? Examples are the folic acid reductase inhibitors, such as aminopterin, the purine and pyrimidine analogs used as cytostatics in cancer chemotherapy and known for their high toxicity in a wide variety of species, and the organic phosphates and carbamates used as insecticides but also highly toxic to mammals. Lack of selectivity in the action of metabolic inhibitors is inherent in their mechanism of action due to the universality of biochemical processes and principles throughout nature. Selectivity in action requires species differences in biochemistry. For the antivitamins, for instance, there is not only a lack of species differences in action; in addition, the fact that vitamins often serve as cofactors for a variety of enzymes is a serious drawback to endeavors to obtain agents with species-selective action.

2.1.2. Antimetabolites with Selectivity in Action

The selectivity in action of certain useful anti-infectious agents, such as the sulfanilamides (para-amino benzoic acid antagonists) and the penicillins, is based on species differences. Para-amino benzoic acid serves as a precursor for folic acid synthesis in micro-organisms, while higher animals use folic acid itself as a vitamin; penicillin interferes with the synthesis of the cell walls which are required by micro-organisms but do not exist in higher animals. There are interesting efforts to develop selective antibacterial agents by C-fluorination of bacterial cell-wall constituents such as D-alanine[73]. In an analogous way selective insecticides may be developed that act by inhibiting chitin synthesis[74]. Analysis of the enzyme systems involved in the anaerobic metabolism of various intestinal parasites and pathogenic micro-organisms, especially the obligate anaerobic ones, opens up prospects for the development of agents that act selectively against such organisms[75, 76].

Although, as mentioned, the biochemical pathways throughout the world of living things are rather similar, a closer analysis does show differences. The enzymes involved in identical biochemical conversions in different species or even tissues, although isodynamic, differ in structure, as can be shown immunologically. They are called iso-enzymes. This situation is comparable to that of the various hemoglobins where the heme moiety, the active site, is identical in different species but the globin moiety differs. These differences very seldom interfere with the binding of oxygen to the hemoglobin. Similarly in isodynamic enzymes of various species or various tissues, although the active sites are practically identical, the areas on the enzyme surface surrounding the active site may differ. Blocking agents that bind not only to the active site, but also to the surrounding accessory binding areas may show species selectivity because of structural differences in these areas. Hitchings, using folic acid reductase of bacterial and mammalian origin, succeeded in developing selective inhibitors on the basis of such differences (Fig. 3)[77, 78].

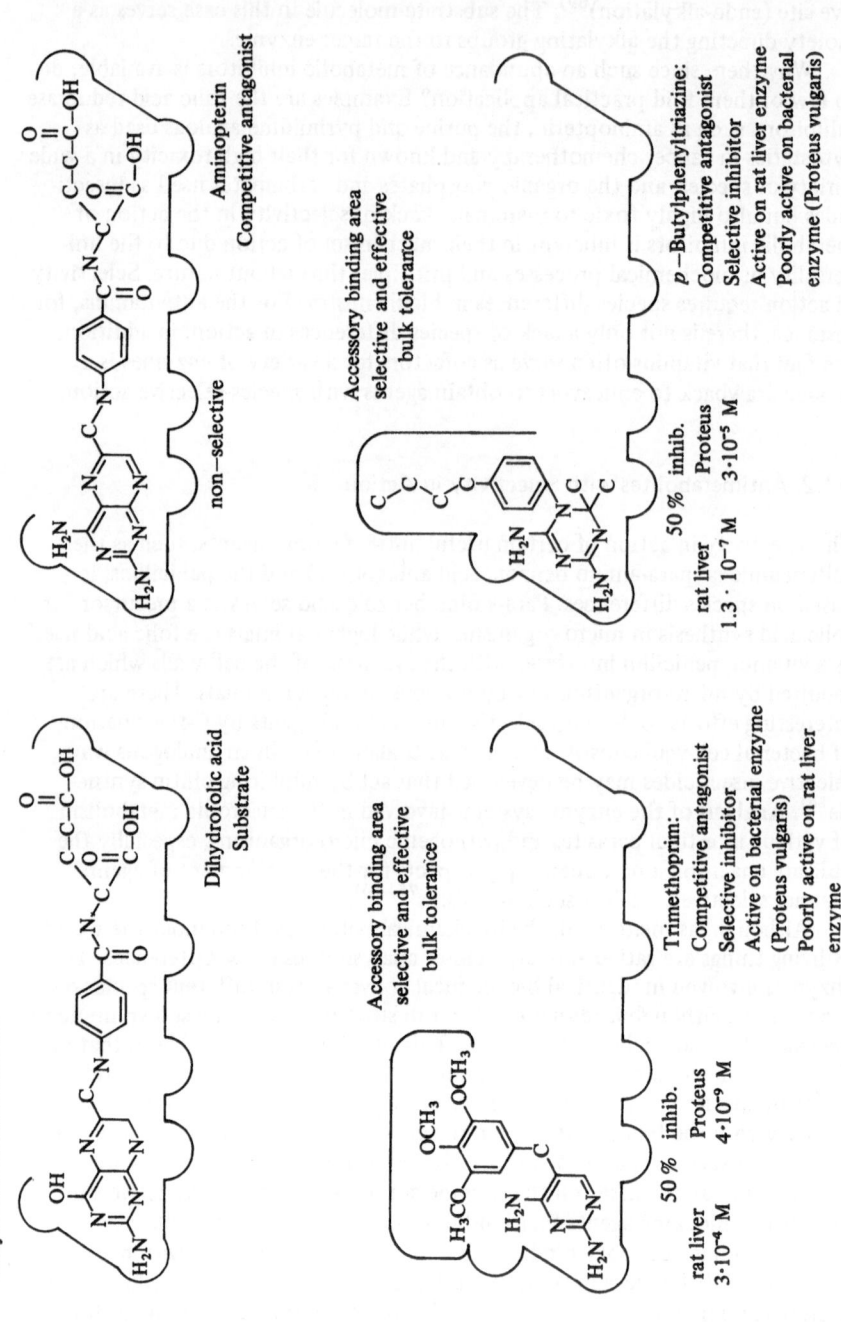

Fig. 3. Design of metabolic inhibitors binding on accessory binding sites to obtain selectivity in action[77]

2.1.3. The Point of Attack in the Metabolic Pathway

Metabolic inhibitors may fail even where no species selectivity is required. The antimetabolite $\Delta_{4,5}$-cholestenone, designed to inhibit cholesterol synthesis, illustrates this: it blocks the conversion of desmosterol to cholesterol, the final step in this pathway (Fig. 4). The blockade of cholesterol formation, however,

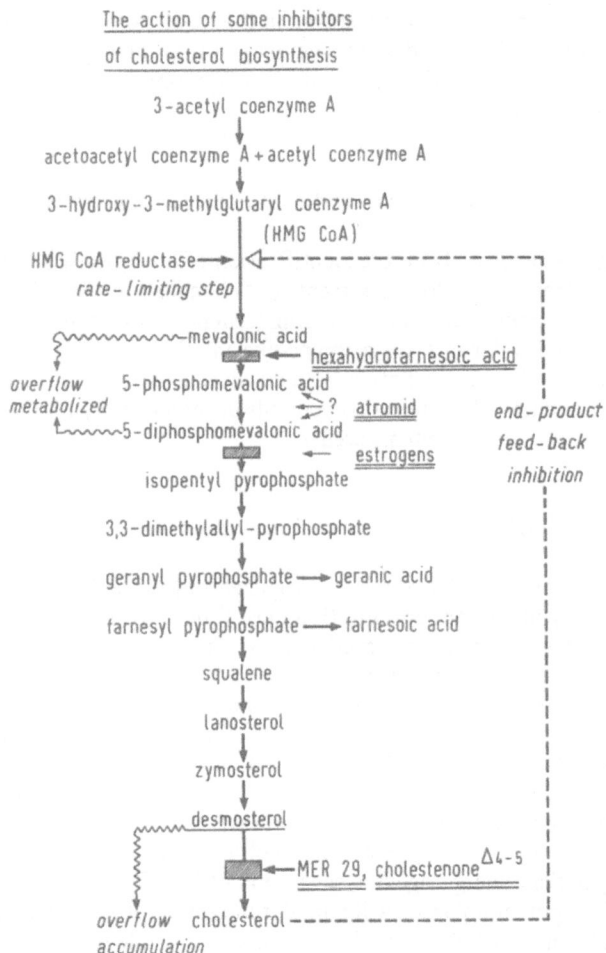

results in elimination of the feedback control which is normally performed by cholesterol[79]. The consequence is unrestricted formation of the intermediate desmosterol, which tends to override the blocking agent and accumulates in the tissues[1, 80, 88].

11

Compounds designed to modulate regulation in metabolic pathways should preferentially act on the enzyme involved in the rate-limiting step, which is as a rule the regulatory enzyme controlled by the feedback mechanisms. This is often the first step after branching in a biochemical pathway. Other potential modulators of biochemical processes, besides blockers of the rate-limiting enzyme, are mimetics or antagonists for the feedback regulators acting on allosteric sites on this enzyme. Here too, the capability to mimic or antagonize repressor and derepressor agents involved in the regulation of enzyme synthesis opens up prospects. The rapidly growing insights into the regulatory processes in biochemistry will point the way to new and appropriate modulators of biochemical pathways.

2.2. Screening of Natural Products and Synthetics for Bioactivity

Many drugs have evolved from the identification of biologically active natural products and from the biological screening of chemicals in general, especially of chemical compounds with new and unprecedented structures, e.g. the various curariform muscle relaxants, the atropine-like anticholinergic agents, and the various local anesthetics originating from cocaine. The degree of success achieved by random screening of chemical compounds depends largely on the relevance and the scope of the test systems used[1].

2.2.1. Separation of Therapeutic Action from Unwanted Action

Once a suitable type of bioactivity is recognized and the active principle has been chemically identified, optimization must follow. This involves among other things elimination of unwanted actions, if present, by chemical manipulation. If the pharmacodynamic basis of the side-effects differs from that of the desired effect, there is a good chance of separating them, since the difference in mechanisms means that the physicochemical properties of the molecule that produce the one and the other effect may differ sufficiently[1]. If the desired effect and the side-effect have the same pharmacodynamic nature, which implies that they are induced by identical pharmacodynamic mechanisms (e.g. the inhibition of salivary secretion, mydriasis, and the inhibition of gastric motility by anticholinergic agents), separation of the effects is problematic. Incidentally, ophthalmologists will consider inhibition of salivary secretion as a side-effect and mydriasis as the therapeutic effect, while for specialists in internal medicine inhibition of salivary secretion may be the therapeutic effect and mydriasis the unwanted side-effect. This underlines the relativity of the terms therapeutic effect and side-effect. In some cases the main effect and the side-effect are induced by different molecules, namely the drug administered and one of its metabolites. Strictly speaking, every drug effect must be regarded as an undesired effect in a healthy patient or in a patient who receives the drug on a wrong indication.

2.2.2. From Side-effect to Therapeutic Effect

The unwanted side-effects of drugs can be a productive source of leads to new compounds. The evaluation of side-effects for their possible use in therapeutics has made a substantial contribution to our therapeutic arsenal. The various phenothiazine-type tranquillizers, for instance, are derived from the originally undesired sedative side-effect of promethazine, a phenothiazine-type antihistamine. Further incidental observations led to a large family of phenothiazines and related drugs (Fig. 5).

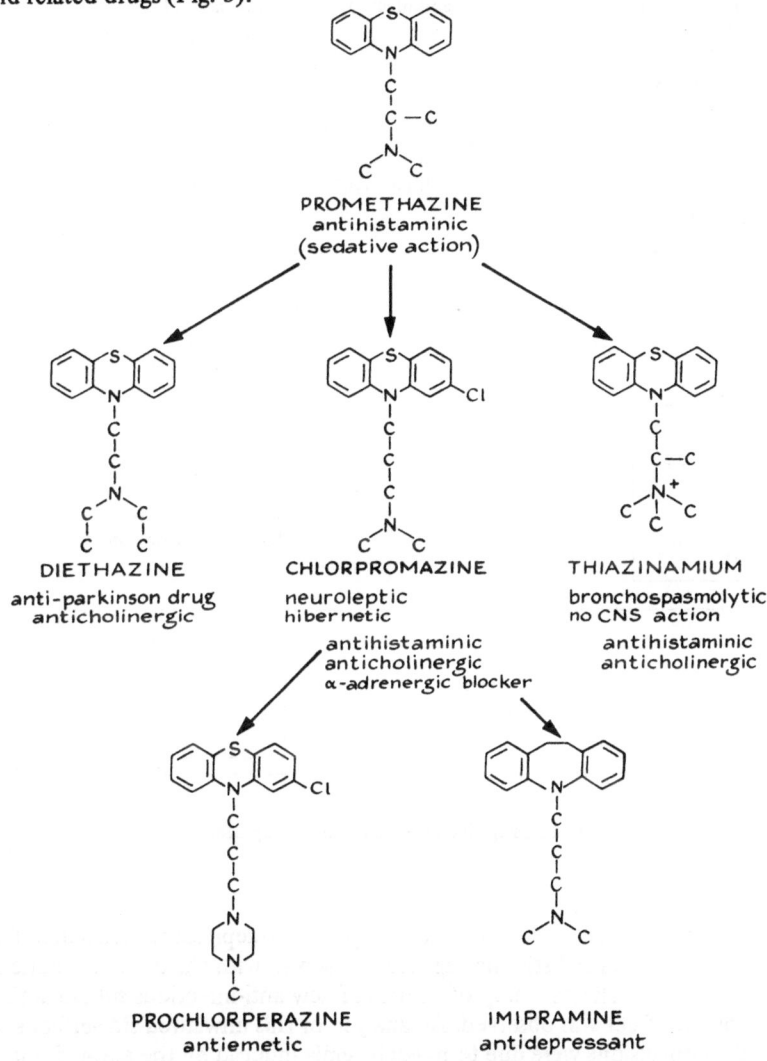

Fig. 5. Genealogy of various psychopharmaca

13

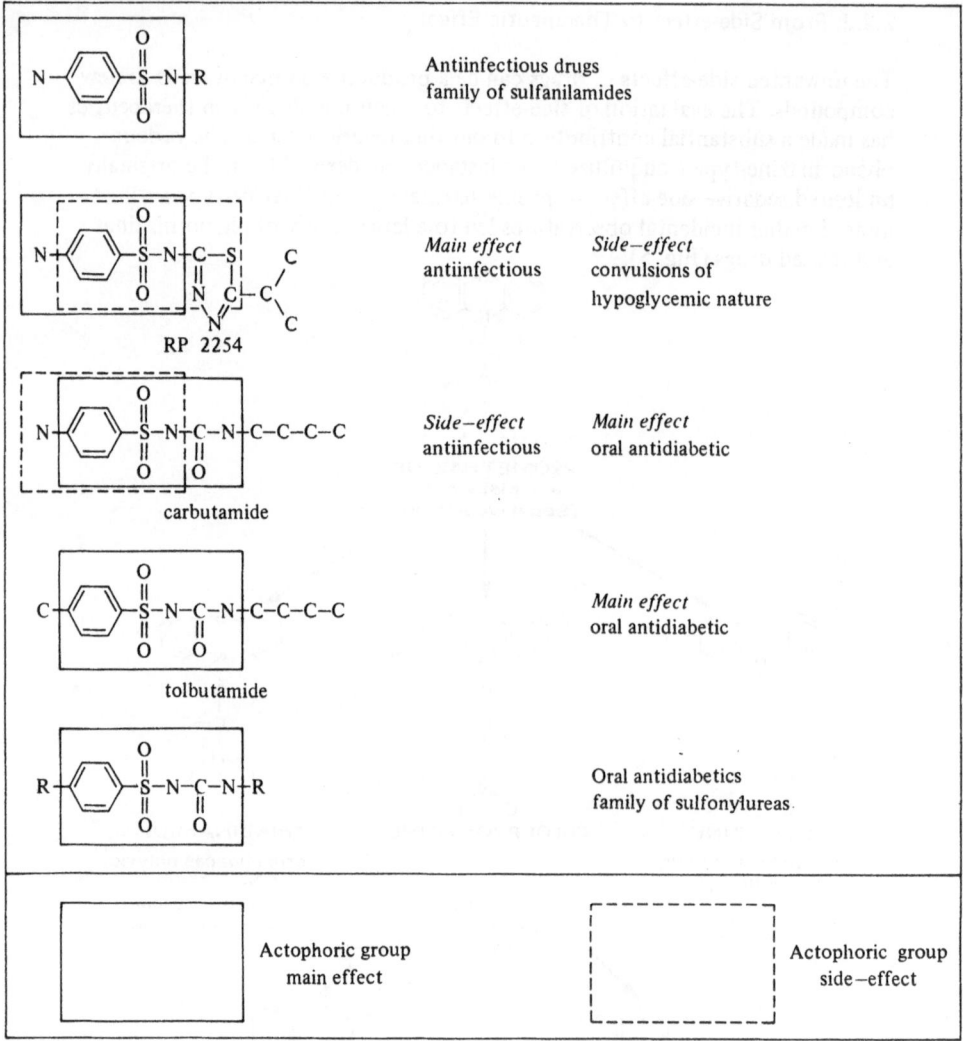

Fig. 6. Drug side-effect as lead in drug design

The monoamine-oxidase inhibiting type of antidepressants originated from the unexpected central-stimulating effect observed with the tuberculostatic iso-niazid. In the preclinical study of a series of new anti-infectious sulfanilamides, a convulsive effect was observed. An analysis of this unwanted side-effect showed that the convulsions were due to hypoglycemia induced by the anti-infectious sulfanilamide. This hypoglycemic action was recognized as potentially useful

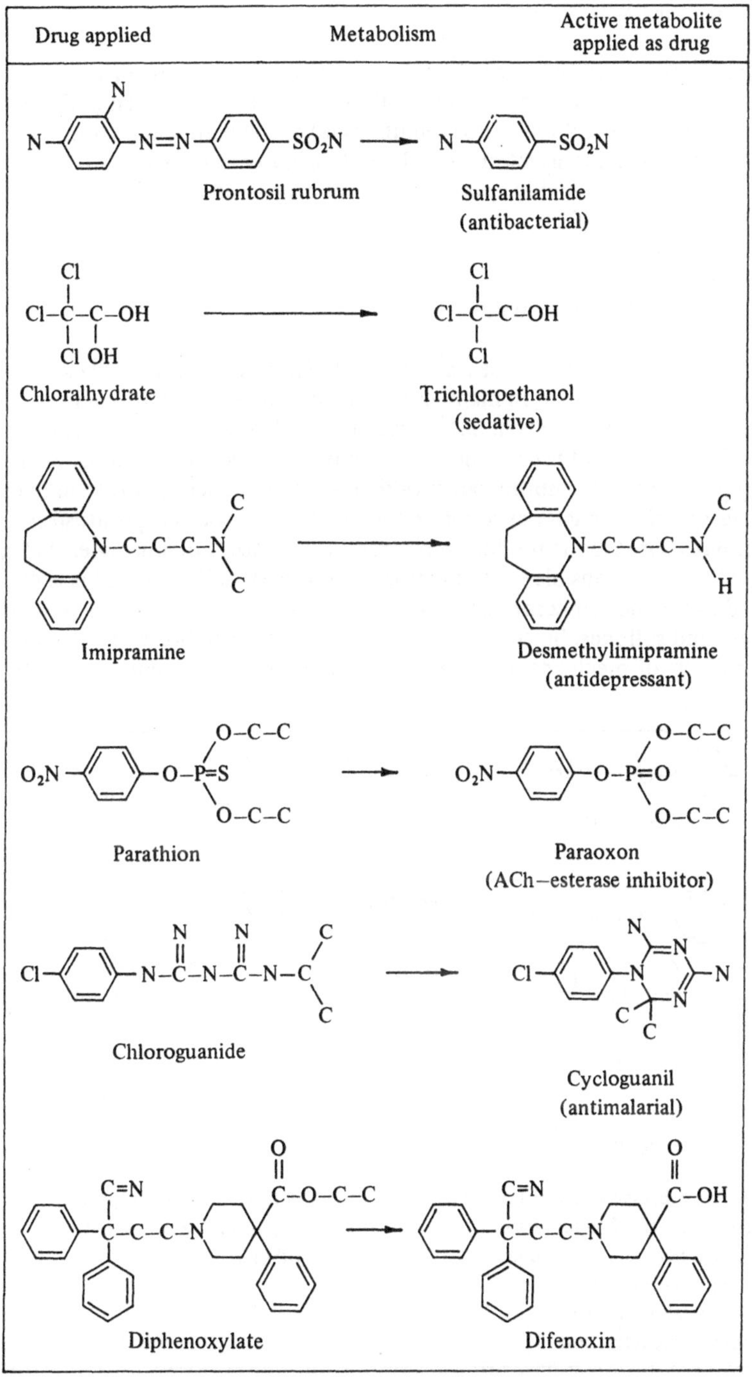

Drug applied	Metabolism	Active metabolite applied as drug

Prontosil rubrum

Sulfanilamide
(antibacterial)

Chloralhydrate

Trichloroethanol
(sedative)

Imipramine

Desmethylimipramine
(antidepressant)

Parathion

Paraoxon
(ACh−esterase inhibitor)

Chloroguanide

Cycloguanil
(antimalarial)

Diphenoxylate

Difenoxin

Fig. 7: Drug metabolites as potential drugs

15

in the treatment of diabetes[83–85]. Via antidiabetics with the original anti-infectious action as a side-action (carbutamide) this lead resulted in the development of pure oral antidiabetics (tolbutamide) and subsequently a range of related, short- and long-acting, weakly and highly potent antidiabetics (Fig. 6)[1]. Various drugs, now highly appreciated, have originated from following up "unwanted" side-effects.

2.3. Bioactive Drug Metabolites

Another interesting source of leads in the design of new drugs is the study of pharmacon metabolism. In many cases it is not, or not only the chemical compound administered as such, but its metabolites that appear to contribute to the effect. The agent used is then in fact a pharmacogen or pro-drug. A classic example is prontosil rubrum, an anti-infectious agent, active *in vivo* in mice but devoid of antibacterial action in the test tube. It was not prontosil rubrum itself but its metabolic product generated in the mice, sulfanilamide, that appeared to be responsible for the anti-infectious action. Sulfanilamide then served as the mother compound for a wide variety of anti-infectious sulfanilamides and sulfones. Since the study of pharmacon metabolism has become a normal part of preclinical and clinical investigation, the phenomenon of bioac-

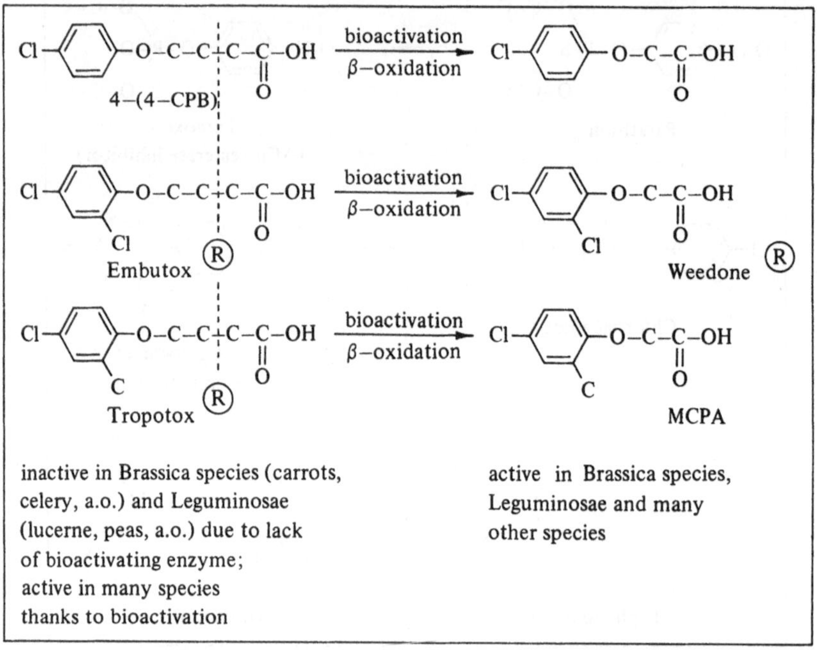
bioactivation
β–oxidation
4–(4–CPB)
Embutox
Weedone
Tropotox
MCPA
inactive in Brassica species (carrots, celery, a.o.) and Leguminosae (lucerne, peas, a.o.) due to lack of bioactivating enzyme; active in many species thanks to bioactivation
active in Brassica species, Leguminosae and many other species

Fig. 8. Selectivity in weedkillers based on selective bioactivation

tivation has been found to be not at all uncommon. Indeed, there are now various drug metabolites used as drugs, or as leads for the design of new drugs (Fig. 7)[1, 86, 87].

Bioactivation can be exploited to obtain selectivity in action. In the field of insecticides, bioactivation in the plant, known as systemic action, is the means used to obtain selectivity in action. Thus, only insects feeding on the sprayed plants will be affected[89, 111]. Similarly, selectivity in the action of weedkillers may be based on selective bioactivation in the target plant[90, 91, 142] (Fig. 8).

2.3.1. Biotoxification

Pharmaca can also be converted metabolically to toxic products in mammals. Special attention must therefore be given to the study of the processes involved in biotoxification since an understanding of these processes may indicate how to avoid toxic action by correct molecular manipulation.

Where toxic effects are based on chemical lesions, the identification of the chemically reactive molecular species involved requires special attention. Pharmaca as drugs, food additives, food contaminants, etc. are relatively stable chemically, otherwise they would decompose before entering the biological object and reaching their sites of action. The biological alkylating agents used as cytostatics in cancer chemotherapy are an exception in this respect. The pharmacon metabolites that are excreted are also relatively stable chemically. Then where do the chemical lesions originate? In pharmacon metabolism, especially in oxidative conversion by the microsomal mixed-function oxidative system, short-lived reactive intermediate products occur, for example the electrophilic epoxides, compounds with biologically alkylating capacities formed in the process of aromatic ring oxidation (Fig. 9)[10, 11, 13, 22, 168, 172]. N-oxides and hydroxylamines are intermediate products in the oxidation of amines. Hydroxylamines may, after conjugation with e.g. sulfates, form reactive electrophilic products with biological alkylating capacity (Fig. 10)[7−9, 11, 12, 15, 170]. The electrophilic intermediate metabolic products tend to bind with critical cell constituents, such as proteins and nucleic acids, and to form covalent bonds by attack on nucleophilic groups such as SH- and in particular NH_2-groups. The action radius of the reactive intermediates depends on their half-life and thus determines the localization of the chemical lesions. Products with highly reactive electrophilic groups will react with the water that is present in abundance.

Species differences with regard to various toxic effects based on such chemical lesions and their localization in particular tissues are understandable on basis of species and tissue differences in the capacity of metabolizing pharmaca. The pharmacon under these circumstances acts as a toxogen (compare pharmacogen). It may be emphasized that, although the microsomal mixed-function oxidative system in the liver cells is generally considered to be a de-

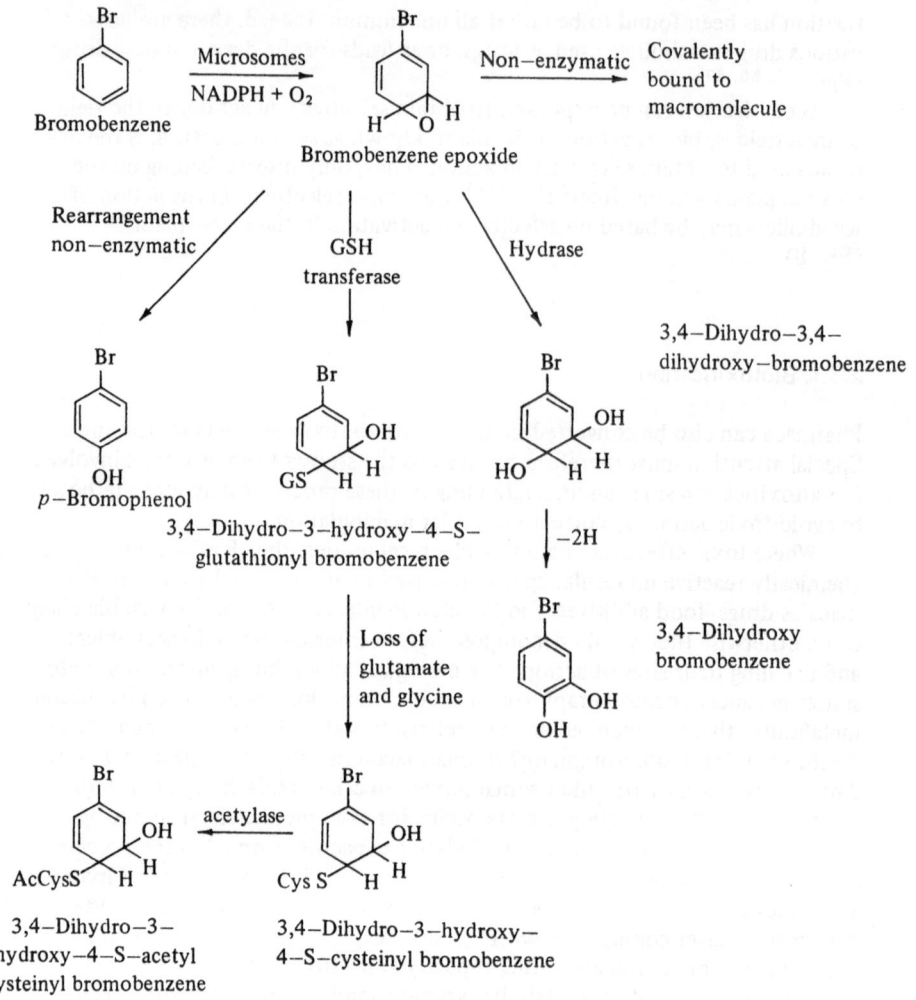

Fig. 9. Pathways of bromobenzene metabolism[22]

toxification system, it too often acts as a toxification system, generating very troublesome toxic effects.

The true detoxification products as a rule appear in the urine, while the reactive intermediate toxic products remain in the body, tightly bound to tissue components, and thus chemically sequestered in the tissues. From the toxicological point of view therefore, it is not the percentage of the pharmacon and its metabolites recovered from urine, feces, and from the carcass by extraction that is most important in balance studies; it is in fact the fraction of the

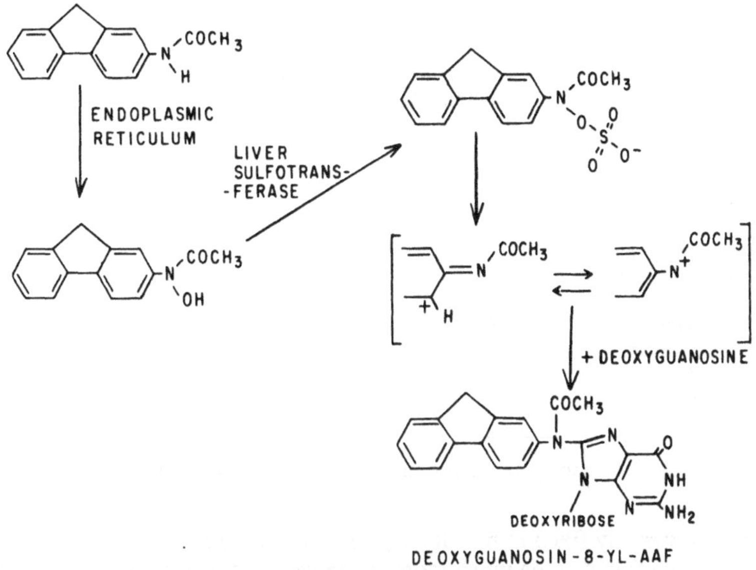

Fig. 10. Pathway of 2-acetylaminofluorene metabolism[11]

compound which is chemically sequestered and therefore cannot be extracted from the tissues. A recovery of 80% whithout any nonextractable residue in the tissues is a better index of the long-term safety of a compound than is a recovery of, say, 95% with a 2% chemically sequestered residue, which points to induction of chemical lesions with all the likely consequences. Simple test systems, such as tissue cultures, and fungal or bacterial cultures to detect possible carcinogenic or mutagenic action of pharmaca and of their metabolic excretion products, therefore, are inadequate. Assuming that pharmacon metabolism occurs at all in such test systems, it will probably differ from that in man and mammals in that no intermediate products, or different ones, will be involved. This means that the information obtained with these test systems is not relevant unless the pharmaca are studied in host-mediated tests. The compound is then given to a mammal (e.g. rat) and the plasma of this animal is subjected to the test systems mentioned (Fig. 11)[18, 92, 93, 97]. Another possibility is to introduce the bacterial or tissue cell culture into the peritoneal cavity of a mammal and to study the influence of the compound on the test system after it is given to the animal. Carcinogenic or mutagenic actions based on the intercalation of polynuclear flat molecules in the DNA strand can be detected by direct application to cultures of tissue or microorganisms, and even with DNA in solution[94], since metabolic conversion is not necessary for this type of action. For a number of highly potent carcinogens, such as the polycyclic hydrocarbons, intercalating capacity is combined with conversion to reactive epoxides; the consequence is that after intercalation the molecules

Pharmacon	150 Cells in mitosis, showing		
	Chromosomal anomalies	Chromosomal breaking	Achromatic gaps
Cyclophosphamide 0.84 mg/ml medium 24 h incubation	0	0	0
Cyclophosphamide bioactivated by rat 1 g/kg i.p. serum obtained after 1 h (0.3 ml serum + 9ml medium) 48 h incubation	70	57	12

Fig. 11. Cytogenetic effects of cyclophosphamide on hela cell culture[18]

become irreversibly fixed in position by covalent bonding. Thus intercalating agents may be found to be much more carcinogenic in host-mediated tests than in direct tests.

The cells involved in pharmacon metabolism, especially liver cells, contain systems that detoxify the reactive intermediate pharmacon metabolites, e.g. the enzymes glutathion transferase, which conjugates the electrophilic reactants with glutathion, and epoxy hydrase, which converts epoxides to diphenols (Fig. 9)[22, 95, 96, 165, 172]. Vitamin C is also a potential protectant in this respect[100]. Highly reactive intermediate products will, as mentioned, react directly with the available water and thus be detoxified. Less reactive reactants may be detoxified by the protective systems mentioned. The more stable intermediates, still electrophilic enough to cause chemical lesions, may reach critical enzymes in the cytoplasm, the DNA in the cell nucleus, or even cells and tissues other than those in which they were generated, and develop their cytotoxic actions there. Besides the chemical characteristics of the intermediates, which determine their reactivity and distribution, the rate of formation is also an important factor, since it determines the extent to which the protecting mechanisms will be capable to control the situation. Pharmaca that hardly cause chemical lesions under normal conditions may do so under circumstances of, for instance, drug-induced enhanced drug metabolism (Fig. 12)[32, 99, 101, 102, 166]. The presence of other agents depleting the glutathion pool or interfering with the action of the enzymes engaged in protection enhances the risk of chemical lesions. If one further takes into account that other agents may interfere with the DNA repair mechanisms, it is readily understood that the chance that a potential carcinogenic action will be detected depends on a variety of factors, and that often the combined action of various co-carcinogens is involved.

As indicated in the introduction, chemical lesions are involved not only in carcinogenic and mutagenic actions, but also in allergic actions, tissue damage,

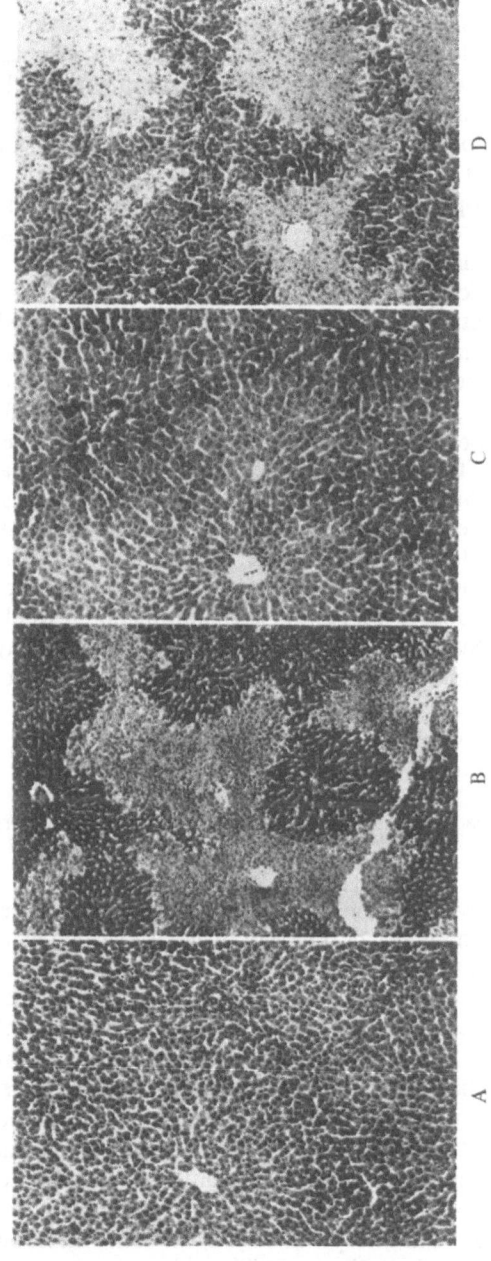

Fig. 12. Paraffin sections of rat liver, periodic acid Schiff stain; X22. *A*, normal liver of control animal killed 24 h after injection of 1 ml of seasame oil i.p. *B*, extensive centrolobular necrosis of parenchymal cells in rat killed 24 h after administration of high dose of bromobenzene (0.2 ml i.p.). *C*, Liver 24 h after lower dose of bromobenzene (0.03 ml i.p.). The centrolobular areas exhibit some small patches of round cell infiltration and decreased glycogen staining in the cytoplasm after hepatocytes but little necrosis. *D*, Extensive centrolobular necrosis after administration of the same low dose of bromobenzene (0.03 ml i.p.) to a rat treated with pheno-barbital (20 mg/kg) for 3 days in order to induce hepatic microsomal enzymes. After Brodie et al.[99]

and local irritation. Design of drugs which are not metabolized via reactive intermediates therefore seems preferable. For control of the type of toxic effects considered, more insight into pharmacon metabolism is needed, especially as regards the formation of reactive intermediates. In studies of chemical lesions caused by radiation, a life-long cumulative tendency is seen, with an exponential increase in risk with total load. There are various reasons why this might also be true for chemical lesions induced by pharmaca. The cytotoxic actions of radiation and of the electrophilic alkylating agents — radiomimetic agents — are related. The "radiomimetic effects" of drug metabolism may indeed have to be added to the known radiation effects. Clearly, this aspect of drug action and hence of drug design calls for special care.

Glutathion acts as a natural protectant against the reactive intermediate products of drug metabolism, which are then excreted as mercapturic acid derivatives[165, 166, 171]. Treatment of experimental animals with high dosages of suitable radioprotectants, e.g. cysteamine (radio-isotope-labeled) before exposure to pharmaca suspected of causing chemical lesions may protect against such lesions[33, 36]. The nucleophilic radioprotectant can act as a scavenger of the electrophilic, biologically alkylating, intermediate products. These will appear then in the urine as conjugates of the radioprotectant fixed to the electrophilic group in the pharmacon metabolite. In this way, information might be obtained on the metabolic conversions leading to the formation of these toxic intermediate products. On basis of such information it would be possible to modify the chemical structure of the pharmaca in such a way as to avoid undesirable metabolic conversions together with the risk of the formation of toxic products. Thus, pharmacon metabolism should always be checked as part of the design procedure.

2.3.2. Half-life of Pharmaca as a Function of Metabolic Inactivation Versus Renal Excretion

Drug metabolism is not only a problem with regard to its involvement in chemical lesions, it is also an important source of uncertainty when predicting drug action in man on the basis of data obtained in animal experiments. This is mainly because of species differences in pharmacon metabolism and consequent differences in the rate and type of bioinactivation (detoxification) and bioactivation (toxification). The same is true of patient-to-patient differences in the dose required for a therapeutic effect, and in sensitivity to toxic side-effects[107–110]. Pharmacon metabolism is also a factor in the complications that arise from drug interactions and the resulting complications of combination therapy[112–115]. This raises the question: "Is drug metabolism essential for drug action?"

If bioactivation is involved in therapeutic action, the active metabolite or a derivative thereof may be used as a drug, possibly with even more success. Termination of drug action and consequent elimination does not necessarily depend on pharmacon metabolism. Drugs are known that are practically not metabolized and yet still have a reasonable half-life and elimination rate. Elimination is determined from renal excretion, especially the degree of passive reabsorption from the renal tubuli. This in turn depends mainly on the partition coefficient and the pK_a of the compound. It follows that adaptation of the half-life of a non-metabolized pharmacon can occur on the basis of adaptation in the partition coefficient (lipid/water solubility) and the pK_a. Even lipid solubility of the degree required for action on the central nervous system is compatible with a reasonable (*e.g.* 24 h) half-life, as can be concluded from such drugs as fenfluramine, pimozide and pempidine. The half-life of many drugs in use to-day is 3—7 h; this enforces a three-doses-a-day regimen, but may well be prolonged to ± 24 h to permit a one-dose-a-day regimen. The majority of drugs are weak acids or bases so that in case of emergency, *e.g.* an overdose, renal elimination can be speeded up by giving sodium bicarbonate (alkalinization) or ammonium chloride (acidification), respectively[116–118].

The question arises as to whether species differences also exist for passive renal excretion processes. Indeed, some more or less systematic and therefore controllable differences do exist. The urinary pH tends to be more acidic in carnivores than in herbivores. Due to differences in the degree of ionization, passive renal reabsorption of weak bases and acids thus will differ for the species mentioned, but in a reasonably predictable way. Also the degree of binding of the pharmacon to plasma albumin will influence the rate of renal excretion. Here too, species differences are reported, but this parameter is accessible to *in vitro* studies. Both protein-binding and passive reabsorption, factors that determine the rate of renal excretion, are related to the partition coefficients of the compounds[119, 120]. The half-life of various non- or poorly metabolized sulfanilamides is strongly dependent on the partition coefficient, as can be seen from Fig. 13[121].

Compound	pK_a	Protein binding (%)	Lipid solubility % drug in ethylene chloride at pH 7.4	Half-life man (h)
		Short-acting		
Sulfathiazole	7.1	77	15.3	4
Sulfafurazole	4.9	86	4.8	6
Sulfisomidine	7.4	86	19.0	7
Sulfaethidole	5.6	99	6.2	7
		Long-acting		
Sulfamethyldiazine	6.7	85	69.6	35
Sulfamethoxypyridazine	7.2	90	70.4	37
Sulfamethoxydiazine	7.0	87	64.0	37
Sulfadimethoxine	6.1	99	78.7	40

Fig. 13. Physicochemical characteristics of sulfanilamides in relation to half-life. After Struller[121]. Only slowly metabolized sulfanilamides are incorporated

In attempting to reduce the risk of chemical lesions by avoiding pharmacon metabolism, one assumes that the rate of elimination by drug metabolism is negligible compared with the rate of elimination by renal excretion processes. Further more, high potency of the compounds means that only small or very small doses are involved and therefore only small or very small quantities of metabolites can be formed, and so reduces the risk of chemical lesions. If short-acting agents, *i.e.* pharmaca with short half-lives, are desired, one might consider introducing moieties highly vulnerable to metabolic attack. A preferentially hydrolytic decomposition of the compounds by plasma or liver esterases should then lead to rapid inactivation, with the formation of metabolically stable and rapidly excreted products. *In vitro* testing of the rate of hydrolysis in plasma or mammalian liver preparations can yield useful information here. If particular pharmacon transport forms are used to facilitate absorption or modulate distribution[2, 49−51], the requirement is for removal of the disposable moieties introduced, usually by hydrolytic cleavage. If compatible with the desired action as such, an increase in water solubility by introduction of *e.g.* strongly ionized moieties into the pharmacon molecule may help to restrict distribution to the extracellular space and thus prevent metabolic conversions.

In practice, the design procedure will always be a kind of compromise seeking to combine the desired biological activity in the target system or species with a proper dose-effect profile and with selectivity in action, especially as regards the avoidance of toxic side-effects.

2.3.3. Avoidance of Biotoxification in the Design Procedure

As discussed above, avoidance of drug metabolism to eliminate chemical lesions and other consequences of drug metabolism is of particular interest in the design of drugs. The fact that certain drugs are broken down into ten or more metabolic products raises the question whether all positions at which the drug molecule is changed have to be protected against metabolic attack. The chance that the many adaptations to the chemical structure that would then be necessary would be compatible with the desired action is vanishingly small. It is becoming more and more apparent, however, that in cases where numerous metabolites are formed, they often originate from one or just a few primary enzymatic attacks on the pharmacon molecule. For instance, in the organic amines, the primary attack is very often on the amino group with formation of an N oxide. Then, secondarily, non-enzymatic intramolecular rearrangements take place in these N-oxide-bearing molecules, resulting in a variety of end products, depending on the structural properties of the molecule concerned[122]. End products that require the lowest activation energy for rearrangement will be most abundant. In such a case the formation of various end products can be avoided by blocking the initial enzymatic oxidation of the amino group. Similar relationships hold for oxidations in aromatic rings, where a number of end products may be formed as a result of only one initial enzymatic oxidative attack, via non-enzymatic rearrangements known as NIH shifts. In the course of the rearrangement processes free radicals and other chemically reactive intermediate products prone to cause chemical lesions can be generated. Avoidance of the initial oxidative step by suitable molecular manipulation implies avoidance of the various sequential chemical conversions. In this respect one might consider protecting *e.g.* phenyl rings against oxidative attack by suitable substitution, for instance of fluorine[164] in *para* position. Allyl groups and other unsaturated moieties suitable for epoxide formation must possibly be considered as suspect and therefore undesirable. A greater problem will be the avoidance of N-hydroxyl product formation. The analysis and systematization of the relationship between chemical structure and pharmacon metabolism, especially with regard to the formation of intermediates having the capacity to cause chemical lesions, must supply the necessary fundamental information.

Pharmaca designed to be resistant to metabolic and especially oxidative attack could be potential persistent environmental pollutants. Since, however, the biological degrading capacity of the water and soil microflora is highly differentiated, compounds stable enough to avoid biotoxification in mammals and other animals may still be sufficiently vulnerable to microbiological attack. Moreover, in the case of therapeutics, food additives and selective, *e.g.* pheromone-derived pesticides only relatively small quantities of substance are involved.

If all drugs, even those that have long been on the market, were tested for their potential contribution to carcinogenesis, mutagenesis, etc. with the newer, more reliable test systems, a fair number of them might be rejected. This would give a strong stimulus to the design of a new generation of therapeutics, namely those having built-in safety with regard to some of the most alarming toxicological risks.

3. Pharmacokinetic Aspects of Drug Design

In the foregoing sections, pharmacon metabolism has been considered in relation to the pharmacodynamic, especially toxic aspects of action. Drug distribution also plays a role in the pharmacodynamic phase of drug action. Modulation of drug distribution, for instance, exclusion of penetration of particular barriers in the organism (*e.g.* the blood-brain barrier), or preferential bioactivation or bioinactivation in particular tissues, can modulate the spectrum of action of compounds and possibly exclude particular undesired effects.

With regard to the pharmacokinetic phase of action, a distinction can be made between distribution over the various compartments or tissues, and the time-concentration relationship in particular compartments. Modulation of the pharmacokinetics by suitable molecular manipulation is important at an early stage in the design procedure. Here a distinction can be made between the modulation of the rate and possibly the site of bioinactivation or bioactivation, and modulation of the distribution, including the rate of absorption and rate of elimination.

3.1. Modulation of Pharmacon Metabolism by Molecular Manipulation

In seeking to control pharmacon metabolism, special attention must be paid to those moieties in the molecule that are particularly vulnerable to pharmacon-metabolizing enzymes: the various hydrolases, the mixed-function oxidases, reductases, etc. The moieties concerned may be called vulnerable moieties. Introduction of such moieties will result in a shortening of the half-life of the compound and hence a shortening of its action. Stabilization or protection of vulnerable moieties in the drug against metabolic attack will prolong the half-life and thus increase the duration of action. The prevalence of particular pharmacon-metabolizing enzyme systems in particular species or particular tissues or compartments may be exploited to obtain compounds that are selectively bioactivated in the target tissue or target species, or are selectively inactivated outside the target tissue or in non-target species. This helps to increase the therapeutic or safety margin. These ideas will be illustrated by a number of examples.

3.1.1. Introduction of Vulnerable Moieties to Shorten the Half-life of Pharmaca

This procedure is followed to obtain short-acting pharmaca and to avoid cumulative environmental pollution. With regard to drugs, the classic examples are

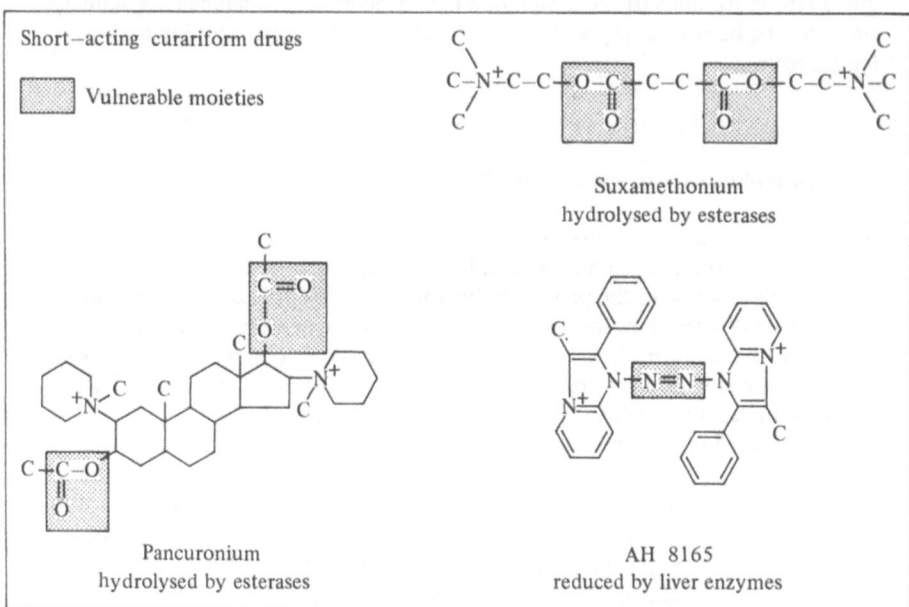

Fig. 14. Introduction of vulnerable moieties to obtain short-acting compounds

the muscle relaxants used in general anesthesia to provide optimal conditions for surgical procedures. The requirements are rapid onset of action to permit immediate intubation and short-lasting action obtained by introduction of a vulnerable moiety, so that the degree of muscular relaxation is easily controlled and any muscle relaxant action is excluded during the postoperative period. In the short-acting muscle relaxants, suxamethonium and pancuronium, carboxy--ester groups are the vulnerable moieties that make the compound sensitive to rapid bioinactivation by plasma esterases. In the short-acting muscle relaxant AH 8165, an azo-moiety vulnerable to reduction by azo-reductase in the liver serves this purpose (Fig. 14)[127–129]. Similarly, there has been a search for intravenous anesthetics with a short duration of action to be used for short-lasting intravenous general anesthesia. A short-acting anesthetic based on rapid redistribution and uptake in body fat has the disadvantage that the recovery period after anesthesia lasts longer. Therefore, ultra-short action based on metabolic degradation is preferred. The ultra-short acting intravenous anesthetic, propanidid, was derived from a slower-acting compound by introduction of a carboxyl ester group; rapid hydrolysis by carboxy-esterases unmasks a free carboxylic group, which ensures rapid inactivation. The same applies to the intravenous anesthetics, etomidate and proxamate (Fig. 15)[123–126].

3.1.2. Switching from "Hard" to "Soft" Pollutants

Introduction of vulnerable moieties is becoming standard procedure for preventing the accumulation of pollutants in biological systems. The classic example

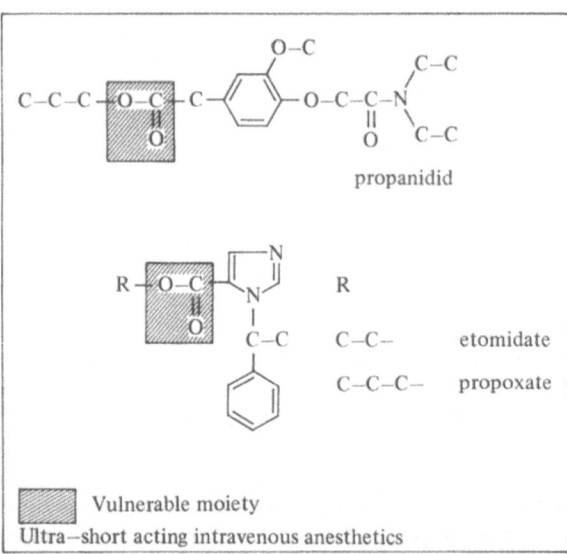

Fig. 15. Introduction of vulnerable moieties to obtain short-acting compounds

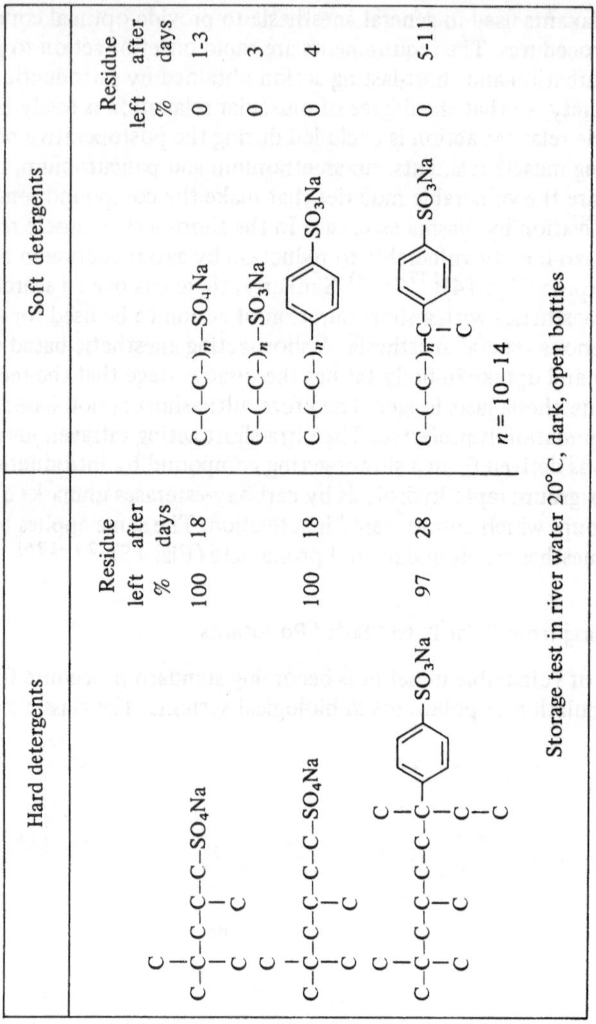

Hard detergents		Soft detergents	
	Residue left after % days		Residue left after % days
$C-C-C-C-C-SO_4Na$ (with branch C)	18 100	$C-(-C-)_n-SO_4Na$	1-3 0
$C-C-C-C-C-SO_4Na$ (branched)	18 100	$C-(-C-)_n-SO_3Na$	3 0
branched SO_3Na benzene structure	28 97	$C-(-C-)_n-$ benzene SO_3Na	4 0
		$C-(-C-)_n-C-C$ benzene SO_3Na	5-11 0
		$n = 10-14$	

Storage test in river water 20°C, dark, open bottles

Fig. 16.[130]

is the "soft" detergents, obtained by substitution of a metabolically vulnerable straight alkyl chain for the biodegradation-resistant branched alkyl chain in the "hard" detergents (Fig. 16)[130]. The insecticide DDT, notorious for its strong tendency to accumulate in biosystems, is highly lipid-soluble and resistant to metabolic degradation; the introduction of suitable vulnerable moieties results in biodegradable insecticides that are therefore less damaging as environmental pollutants. A complicating factor here is the resistance of various insect strains to DDT, due to their enhanced capacity for dehydrochlorination by which the insecticide is converted to an inactive but, as far as environmental pollution is

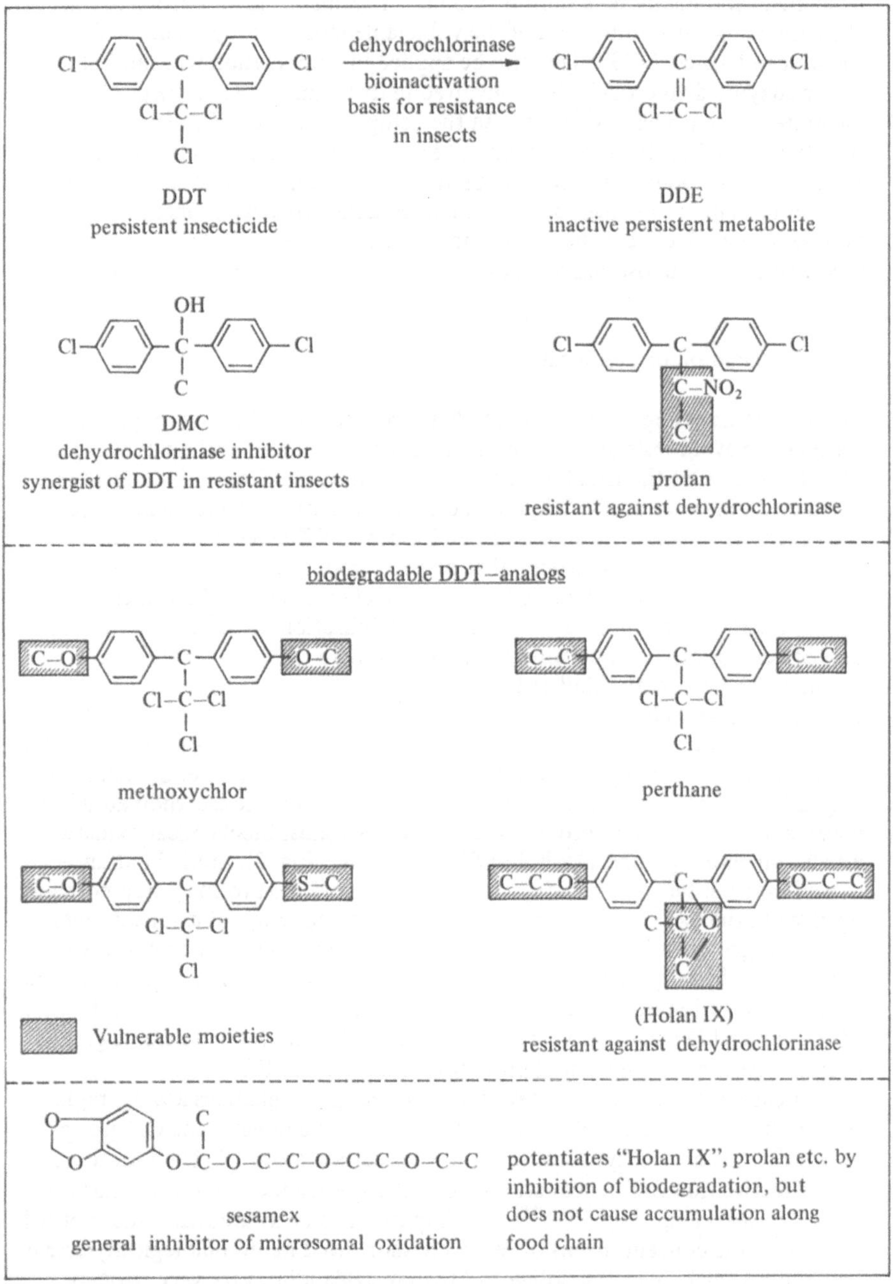

Fig. 17. Design of functional DDT-analogs

concerned, still highly persistent metabolite. The answer to this problem lies in the elimination or stabilization of the vulnerable moiety involved in the development of resistance. DDT might also be applied in combination with inhibitors of the enzyme dehydrochlorinase involved in its bioinactivation. Figure 17 illustrates these principles[131, 132]. In the compound Holan IX the vulnerable moiety involved in the development of resistance is eliminated and vulnerable moieties are introduced to prevent cumulative environmental pollution. This compound is now so vulnerable to the biodegrading capacity of insects that, in order to be effective, it has to be combined with a general inhibitor of insect mixed-function microsomal oxidation systems, such as the compound sesamex.

3.1.3. Elimination or Stabilization of Vulnerable Moieties

This procedure is applied in order to obtain compounds with a more prolonged action or to avoid resistance based on rapid biodegradation in the target object. The elimination of the trichloromethane group from DDT to exclude resistance based on an increase in the dehydrochlorinase capacity of the resistant insects is an example of elimination of a vulnerable moiety. The resistance of the penicillins to penicillinase, an enzyme that plays an important role in the resistance of staphylococci to penicillins, is based not on elimination of the vulnerable amide function in the β-lactam ring, but on stabilization of this configuration. This effect was obtained by introducing substituents into the penicillin side-chain, which apparently hinders the enzyme penicillinase in its approach to the vulnerable moiety (Fig. 18)[136].

A rather common procedure is to introduce, close to the moieties vulnerable to biochemical attack by hydrolases, substituents that induce steric hindrance (Fig. 19)[133–135]. In a similar way alkyl substitution on the α-carbon atom appears to be effective in protecting compounds against biochemical oxidative attacks on amino groups, alcoholic OH groups, etc. (Fig. 2). Steric hindrance, as well as the absence of protons on the carbon atom next to the group normally involved in the oxidation, can play a role here. Introduction of α-substituents often changes the substrate (metabolite) to an antimetabolite, a competitive inhibitor of the enzyme converting the substrate concerned. Fig. 2 gives examples of the stabilization of vulnerable moieties by α-methyl substitution. Substitution of the vulnerable *para*-methyl group in oral antidiabetics by more stable groups (Fig. 20) resulted in substantial prolongation of the half-life.

Of current interest are endeavours to protect prostaglandins against rapid bioinactivation through stabilization of the vulnerable moiety, the OH group in position 15, by methyl substitution on the carbon atom 15. The procedure is so effective that differences are found in the spectrum of actions of the two optical isomers (Fig. 21)[137–141]. It is important to be aware that under normal conditions the concentrations of relatively short-lived natural biologically active compounds, such as acetylcholine and the prostaglandins, vary very strongly with the state of activity of the organism. Substitution for the natural products of more stable mimetic agents cannot reproduce the natural harmony achieved

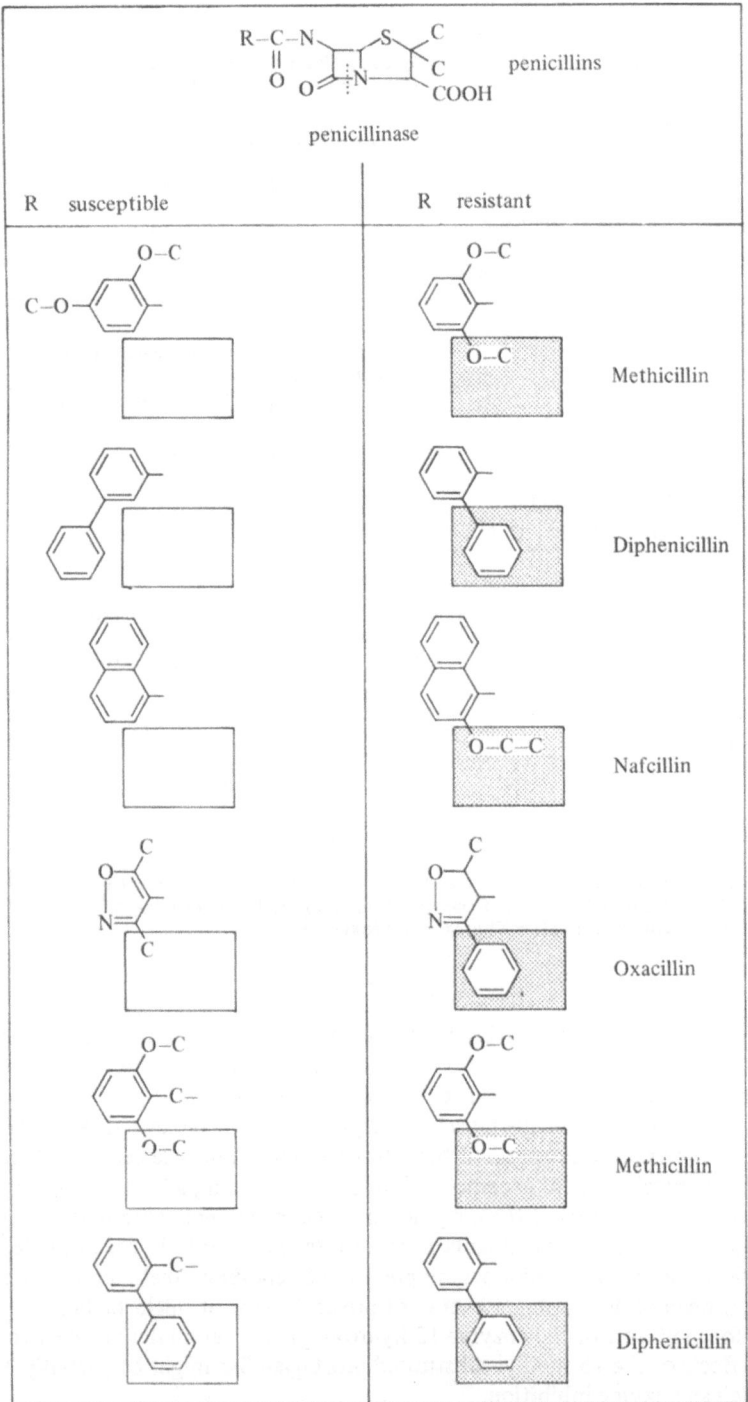

Fig. 18. Structures of penicillinase-susceptible and -resistant penicillins

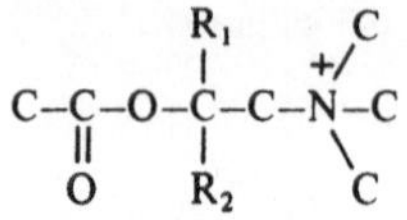

R_1	C

(structure a)

Introduction of stabilizing moieties		Relative rate of hydrolysis (acetylcholinesterase)
R_1	R_2	
Acetylcholine		
$-H$	$-H$	100
L(+)−acetyl−β−methylcholine $-C$	$-H$	54.5
D(−)−acetyl−β−methylcholine $-H$	$-C$	Inhibitor

(a)

Introduction of stabilizing moieties		Relative rate of hydrolysis (human serum)
R_1	R_2	
$-H$	$-H$	500
$-H$	$-C$	15
$-C$	$-C$	0

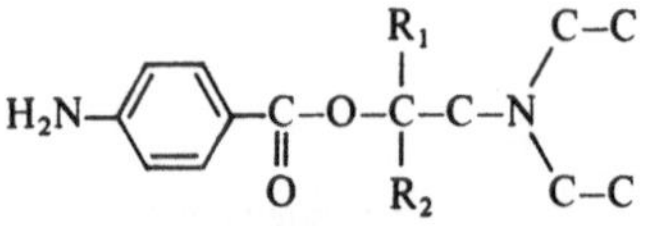

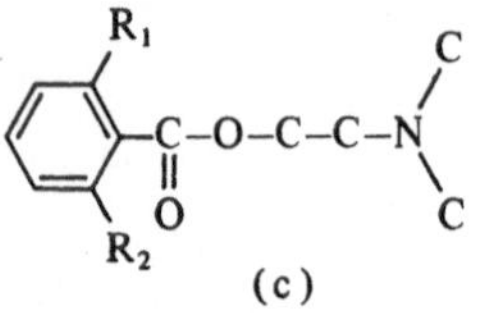

(b)

Introduction of stabilizing moieties		Relative rate of hydrolysis (horse serum)
R_1	R_2	
$-H$	$-H$	100
$-H$	$-C$	65
$-C$	$-C$	0

(c)

Fig. 19. Stabilization of vulnerable moieties by alkyl substitution adjacent to the vulnerable moiety, also called packing of the vulnerable moiety. (a) After Beckett[133]; (b) after Levine and Clark[134]; (c) after Thomas and Stoker[135]

by the continuously fluctuating release of short-lived natural compounds. Substitution for acetylcholine of acetylcholinomimetics (*e.g.* decamethonium or suxamethonium) will not enhance normal muscular function; on the contrary, after initial twitching of the muscle fibers, the muscle is paralyzed. With methacholine, contraction or spasm of the intestinal smooth muscle can be induced, but not peristalsis. If enhancement of function is the aim, a better approach is to introduce inhibitors of the enzymes involved in the rapid elimination of the natural substances. In myasthenia gravis or intestinal paralysis, for example, acetylcholine-esterase inhibitors are preferred to cholinomimetic agents. Similarly, instead of developing mimetics of prostaglandins, it might be better to look for inhibitors of the enzyme 15-hydroxy-prostaglandin dehydrogenase. The effects of the 15-methyl-substituted prostaglandins might be partially based on such an enzyme inhibition.

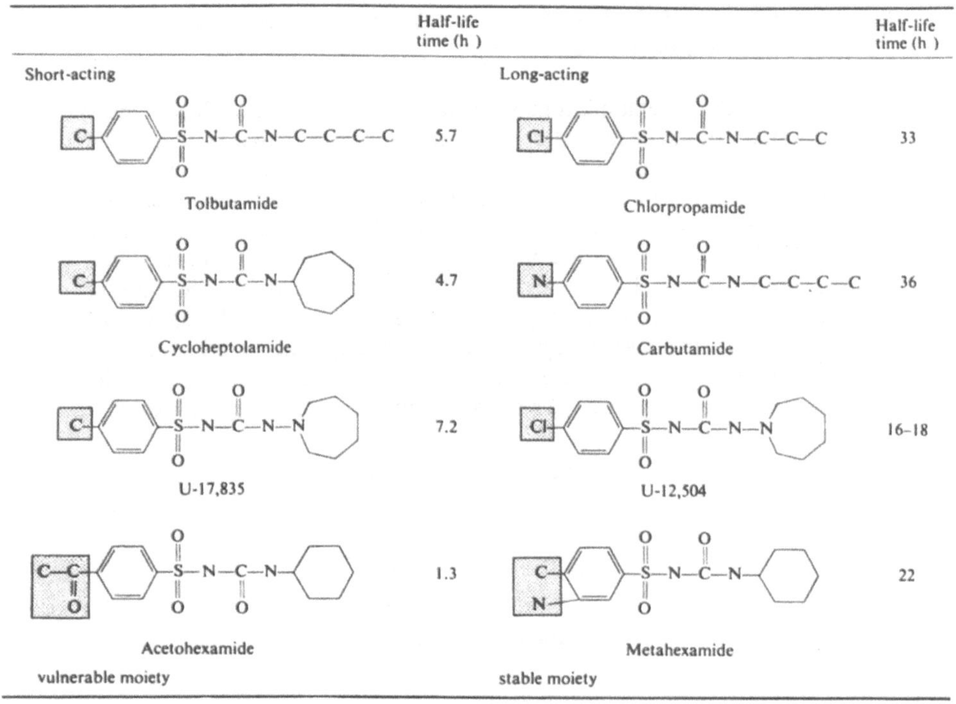

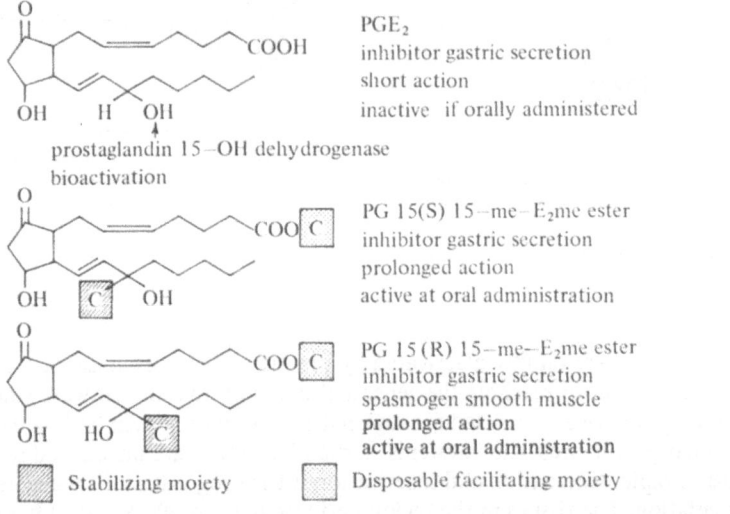

	Half-life time (h)		Half-life time (h)

Short-acting

Tolbutamide   5.7

Cycloheptolamide   4.7

U-17,835   7.2

Acetohexamide   1.3

vulnerable moiety

Long-acting

Chlorpropamide   33

Carbutamide   36

U-12,504   16–18

Metahexamide   22

stable moiety

Fig. 20. Conversion of short-acting oral antidiabetics to longer acting compounds by substitution of the vulnerable moieties by stable groups

PGE_2
inhibitor gastric secretion
short action
inactive if orally administered

prostaglandin 15–OH dehydrogenase
bioactivation

PG 15(S) 15–me–E_2me ester
inhibitor gastric secretion
prolonged action
active at oral administration

PG 15(R) 15–me–E_2me ester
inhibitor gastric secretion
spasmogen smooth muscle
prolonged action
active at oral administration

Stabilizing moiety   Disposable facilitating moiety

Fig. 21. Introduction of a stabilizing moiety and a facilitating moiety in prostaglandin PGE_2.
[From: Karim, S. M. M. et al.[137].]

3.1.4. Introduction of Moieties Suitable for Selective Bioinactivation or Selective Bioactivation

As mentioned, drug metabolism may result in a loss of action with consequent bioinactivation, or in the formation of bioactive, possibly toxic products. Differences in the metabolizing capacity of various organs or of various species, possibly based on differences in the enzyme content, can serve as a basis to obtain selectivity in action. Local anesthetics should act locally, namely at the site of injection, by blocking the propagation of nerve pulses there. However, if these compounds reach the central nervous system in active concentrations, they cause toxic symptoms, mainly convulsions. This is especially likely when such compounds are abusively injected intravascularly. The introduction of a carboxy ester group (see tolycaine, Fig. 22) or destabilization of the amide group (see butanilicaine, Fig. 22) results in compounds which are less toxic

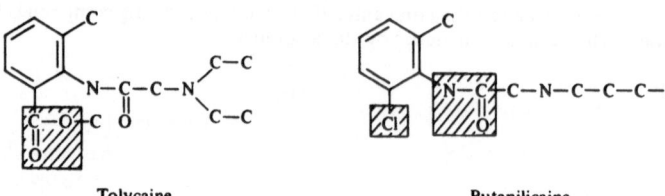

Lidocaine
stable local anesthetic

Tolycaine
introduction of the vulnerable
moiety implies rapid bioinactivation
once general circulation is reached

Butanilicaine
destabilization of the acylamide group
implies rapid bioinactivation once
general circulation is reached

Fig. 22. Introduction of vulnerable moieties to avoid systemic actions in drugs locally applied

because they are rapidly inactivated by plasma and liver esterases and amidases, respectively, after reaching the general circulation. In a similar way, where cytostatic agents have to be used for regional perfusion of certain body areas by extracorporeal circulation systems, carbamide groups are introduced so as to ensure rapid degradation and inactivation if the drug leaks into the general circulation. The tissues in the region perfused (*e.g.* the extremities) have a very low amidase capacity; thus, the general toxicity of the cytostatic is greatly reduced.

The capacity of carboxy esterases and amidases to hydrolyze carboxy esters and amides, respectively, is higher in mammals than in insects. This fact formed the basis for the development of selective insecticides of the organophosphate type. Organophosphates are inhibitors of acetylcholine esterase, an enzyme essential for the functioning of the nervous system, both in mammals and insects. If suitable carboxy ester or carboxy amide groups are introduced into the organophosphates, the resulting compounds are rapidly hydrolyzed in mammalian tissues and hence inactivated, whereas in insect tissues inactivation takes place rather slowly. This ensures rapid detoxification in the economic species, man and higher animals, and thus selectivity in action (Fig. 23)[89].

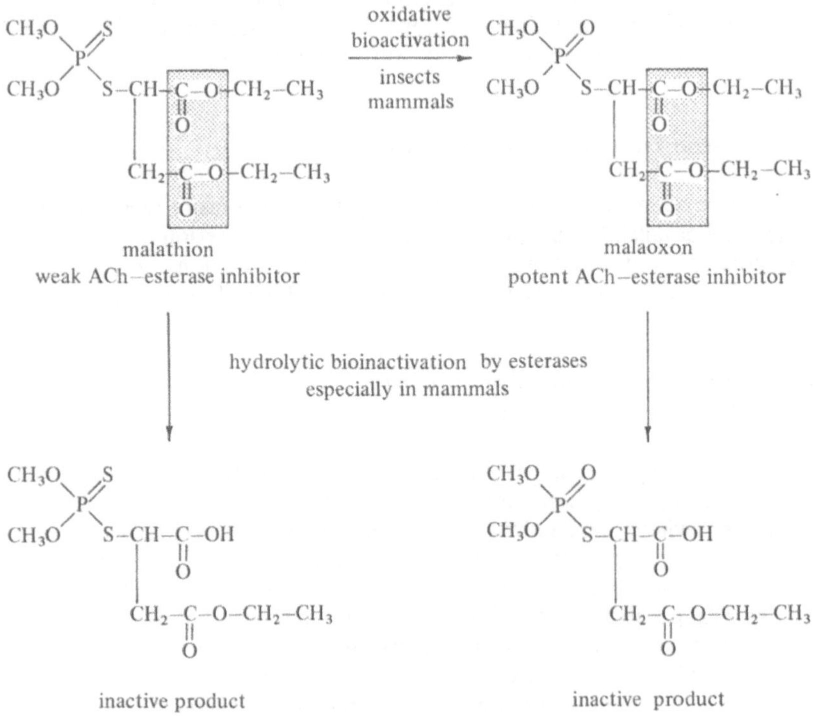

Fig. 23. Malathion: highly toxic to insects, less toxic to mammals. Note: the high capacity to hydrolyse the carboxy ester moiety in mammals as compared to that in insects results in a selective toxicity for the latter

The reverse situation — selective bioactivation in the uneconomic species — is realized in the selectively acting weedkillers of the phenoxybutyric acid type (Fig. 8). The butyric acid derivatives, as such, are inactive; they have to be converted by β-oxidation to the corresponding phenoxyacetic acid derivatives, which have the required auxin and hence weedkiller action. Since plant species differ in their capacity for this β-oxidation, selectivity in action can be obtained. Brassica species, *e.g.* carrots and celery, and leguminosae, *e.g.* peas and lucerne, have a very low β-oxidative capacity and are therefore not sensitive to these weedkillers, while various species of weeds are killed because they rapidly convert the butyric acid derivatives to the active compounds[90, 91, 142].

Efforts to obtain tumor-selective cytostatic agents, based on a selective bioactivation in the malignant tissues, have not been very successful up to now. This is mainly due to a lack of enzyme specificity in malignant tissues as compared to healthy vital tissues. Diethylstilbestrol, an estrogen, is active against tumors of the prostate. The phosphate ester of diethylstilbestrol, Honvan, is inactive but is converted to the active product by the acidic phosphatase that is abundant in the prostate tissue. The enzyme concerned, however, is not restricted to the prostate: kidney and bone tissues are also rich in acidic phosphatase, so that the bioactivation is not restricted to the target tissue.

3.2. Pharmacon Transport and Drug Design

Besides pharmacon metabolism, pharmacon transport is a major factor in determining the bioavailability profile and hence the biological action of pharmaca. Thus, modulation of drug transport by molecular manipulation also offers opportunities to modify the action profile, *e.g.* the half-life of bioactive compounds.

3.2.1. Introduction of Transport-restricting Moieties to Reduce Toxicity

Restriction of drug distribution — for instance, by restricting the penetration of the drug into the compartment where biotoxification takes place, or into the compartment where toxic actions are induced — also opens up prospects for reducing toxicity. Generally speaking, reduction in the absorption, enhancement of renal excretion, and restriction of the distribution to the extracellular fluid, thus hindering penetration into the cells, should reduce the risk both of the formation of toxic metabolites and of intracellular toxic actions. For this purpose, the introduction of highly ionized groups into the molecule is a suitable procedure. Only a small proportion of the highly ionized compounds thus obtained will be absorbed from the gut; once absorbed, they will stay mainly in the extracellular fluid and, after ultrafiltration into the urine, they will not be reabsorbed in the renal tubules. Such compounds are thus rapidly excreted and, since they do not or hardly penetrate the lipid barriers in the various biological membranes, their distribution is so restricted that they do not penetrate into the central nervous system and the intracellular compartments. Certain organs have systems for the active excretion of ionized compounds: the liver and kidneys, for instance, have systems for the active excretion of various strong-acid and strong-base type compounds. These compounds are taken up in a selective way by the cells concerned and rapidly excreted with the urine or bile. Compounds

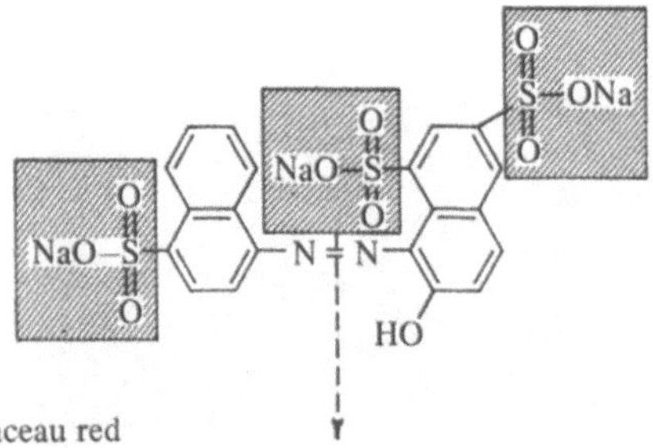
NaO–S
NaO–S
S–ONa
N = N
HO

ponceau red
C.I. food red
safe colorant

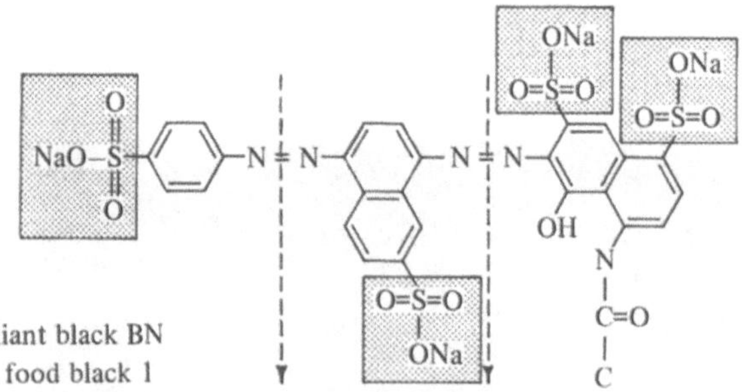
NaO–S
N = N
N = N
ONa
O=S=O
OH
N
C=O
C

brilliant black BN
C.I. food black 1
safe colorant
all split products strongly hydrophilic

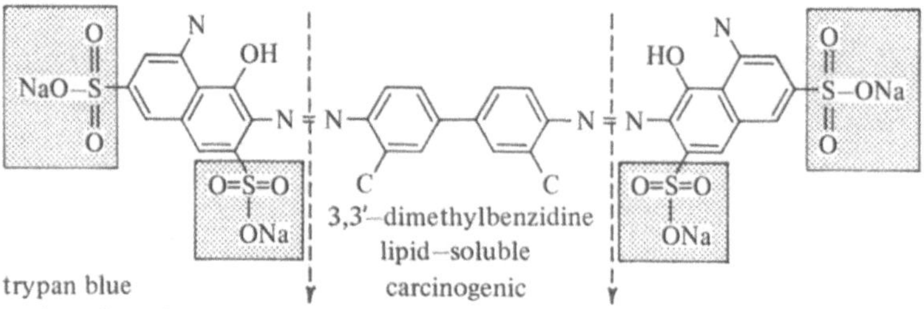
NaO–S
OH
N
N = N
N = N
HO
S–ONa
O=S=O
ONa
3,3'–dimethylbenzidine
lipid–soluble
carcinogenic

trypan blue
toxic colorant

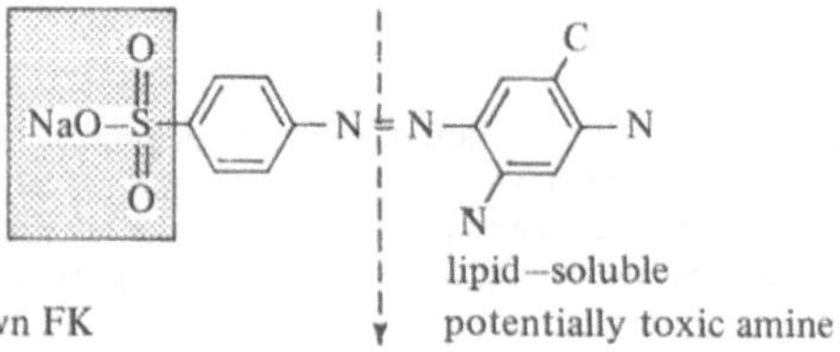
NaO–S
N = N
C
N
lipid–soluble
potentially toxic amine

brown FK
suspicious fish colorant

Fig. 24. Introduction of strongly hydrophilizing moieties to restrict absorption and distribution and to enhance excretion

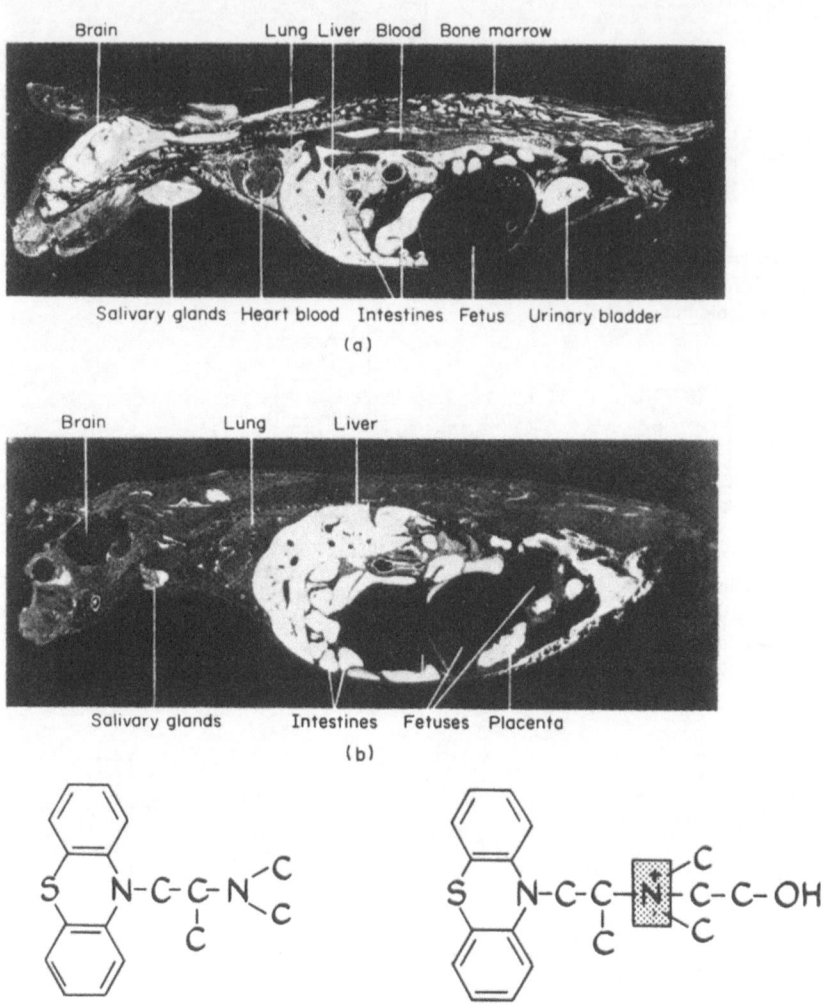

Fig. 25. A comparison of the distribution and excretion of two antihistamines, the tertiary promethazine (*a*) and the quaternary (Aprobit) phenothiazine compound (*b*) labeled with [35]S. Note the high concentrations of the tertiary amine in brain tissue and the restriction in the distribution of the quaternary compound mainly to the liver and the intestines. After Hansson and Schmiterlöw[145)]

with highly ionized groups are as a rule excreted unchanged, which means that they escape attack by the mixed-function oxidases, the enzymes primarily involved in the formation of toxic metabolites. Introduction of onium groups is a less suitable procedure for restricting the distribution, since many cells, such as muscle and nerve cells, have on their surface sites of action (receptors) for quaternary onium compounds, which include acetylcholine and drugs such as curare, hexamethonium, quaternary anticholinergic agents and quaternary antihistaminics. Strongly acidic groups like sulfonic acid groups are preferable, since their presence does not confer any particular pharmacodynamic properties. Examples of the use of sulfonic acid groups as moieties to restrict distribution are found especially in the field of colorants used in foods. Azo dyes, such as

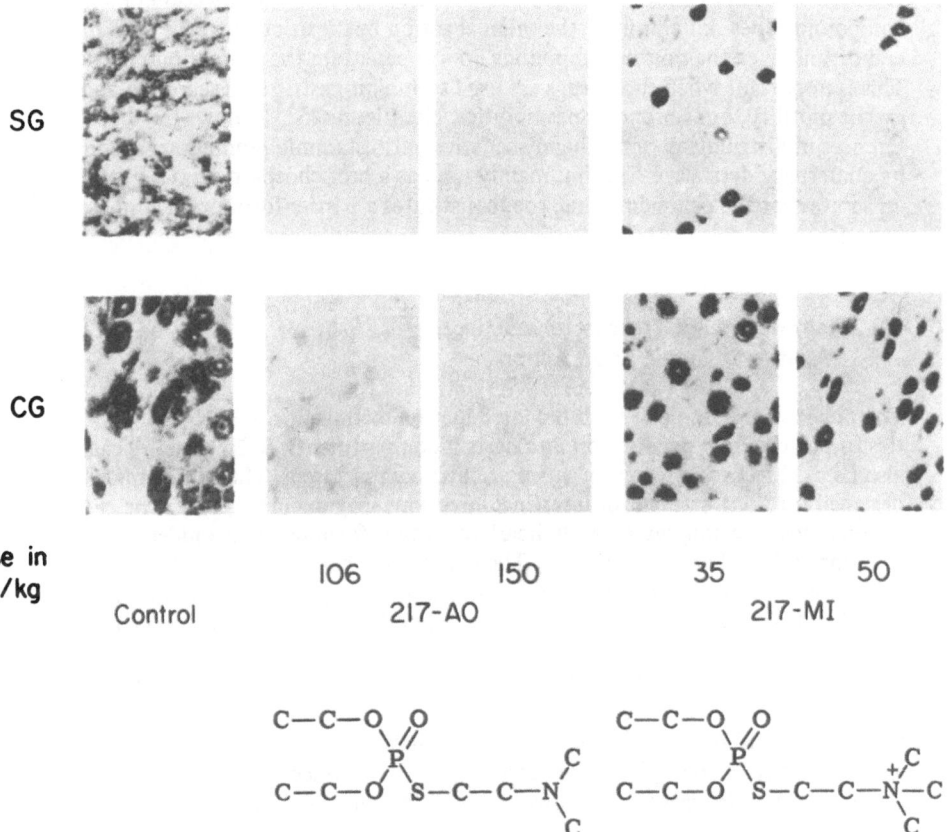

Fig. 26. The penetration of the acetylcholinesterase inhibitors 217-AO (tertiary base) and 217-MI (quaternary onium compound) in interstitial space and cells in the stellate ganglia (SG) and ciliary ganglia (CG) of cats. The preparations are stained on acetylcholinesterase activity. Note: The tertiary compound blocks the enzyme in the extracellular and intracellular space; the quaternary onium compound is restricted in its action to the extracellular space. After McIsaac and Koelle[146)

39

butter yellow, were originally used but were found to be highly toxic, *i.e.* carcinogenic. Among the azo dyes and other dyes accepted as food colorants, highly ionized sulfonic groups are very common (Fig. 24)[143, 144]. One has to be aware that azo dyes, although highly water-soluble, may be reduced to amines particularly by the intestinal flora. Therefore, in these dyes all groups linked by azo bridges should carry at least one sulfonic acid moiety, otherwise after reduction *e.g.* of trypan blue or the suspicious fish colorant, brown FK, toxic and potentially carcinogenic amines may be generated (Fig. 24). Highly ionized groups serving as moieties to restrict distribution can also be used to prevent penetration into the central nervous system so that there are no central nervous system actions. This holds especially true for drugs whose sites of action are on the cell surface, such as anticholinergics and antihistaminics. Quaternization of such compounds does not eliminate the original action but restricts it to the peripheral organs since the onium compounds poorly penetrate the blood-brain barrier. This is important when such drugs are used to inhibit gastric acid secretion and gastric motility, or as bronchospasmolytics, etc. Figure 25[145] shows the differences in distribution of the highly sedative antihistaminic promethazine and its quaternary derivative Aprobit, mainly used as a bronchospasmolytic. Figure 26 shows that the cell membrane, too, constitutes a barrier for ionized compounds[146].

3.2.2. Introduction of Transport-restricting Moieties to Obtain Local Action

The previous section dealt with the introduction of transport-restricting moieties to prevent drug penetration into certain compartments. Such moieties can also be used to promote action in a particular compartment. The sulfonamides designed for the treatment of intestinal infections may exemplify this. These sulfonamides are conjugated with dicarbonic acids, forming hemi-amides, so that one carboxyl group is left free. The conjugates therefore have a strongly acidic character, hence a high degree of ionization and poor absorption in the intestinal tract, allowing the compound to reach the lower parts of the tract (Fig. 27).

For the drug to be antibacterially effective, however, the active sulfonamide must be liberated by deconjugation and the transport-restricting moiety must be disposable. In the intestinal contents slow hydrolysis of the sulfonamide conjugates takes place, thus ensuring effective concentrations of the active antibacterial agent in the lower parts of the intestinal tract.

3.2.3. Introduction of Transport-restricting Moieties to Obtain Selectivity in Action

The principle of the use of disposable transport-restricting moieties is also applied to bioactive compounds in other fields, *e.g.* pesticides. Figure 28 shows

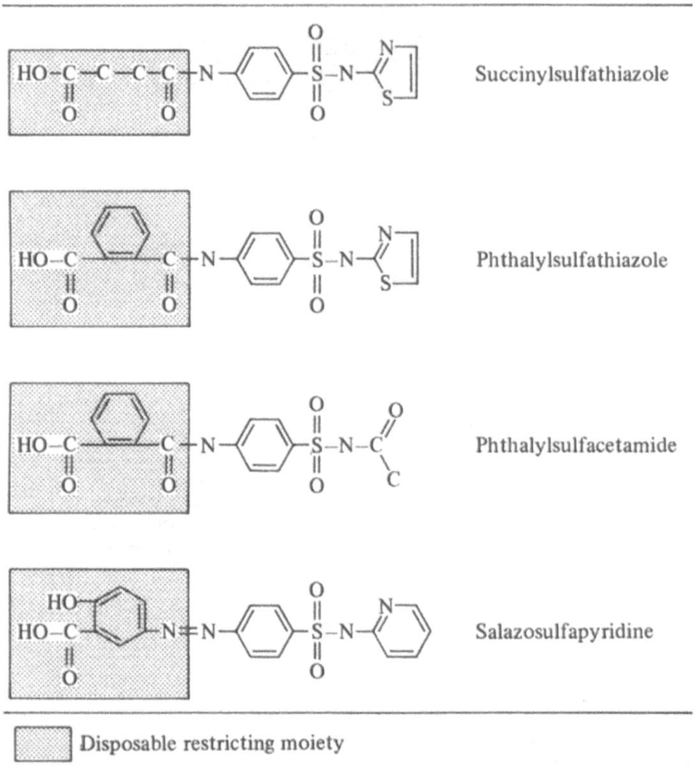

Disposable restricting moiety

Fig. 27. Restriction in the distribution of sulfonamides to the intestinal tract by introduction of strongly hydrophilic disposable restricting moieties

the highly ionized water-soluble conjugation product of dichlorophenoxyethanol with sulfuric acid, a weedkiller which, in this form, does not penetrate the leaves on which it is sprayed. After wash-down by rain the conjugate is hydrolyzed and the alcohol is oxidized to the active product, dichlorophenoxyacetic acid, in the superficial layers of the soil. The action of this weedkiller is thus mainly restricted to weeds having a superficial root system. Because of dilution, it does not touch the deeper-rooted plants such as fruit bushes and trees, even although these have been sprayed with the weedkiller[91]. This procedure can also be regarded as selective bioactivation in the target compartment, the superficial soil layers. It is not only by introducing highly ionized, water-soluble moieties that transport can be restricted, but also by moieties that strongly increase the lipid solubility of the compound so that it is no longer water-soluble. The groups introduced to enhance the lipid solubility can also be regarded as transport-restricting and may also be disposable. This is the case, for instance, for the esters of the radiopaque agents used in bronchography

41

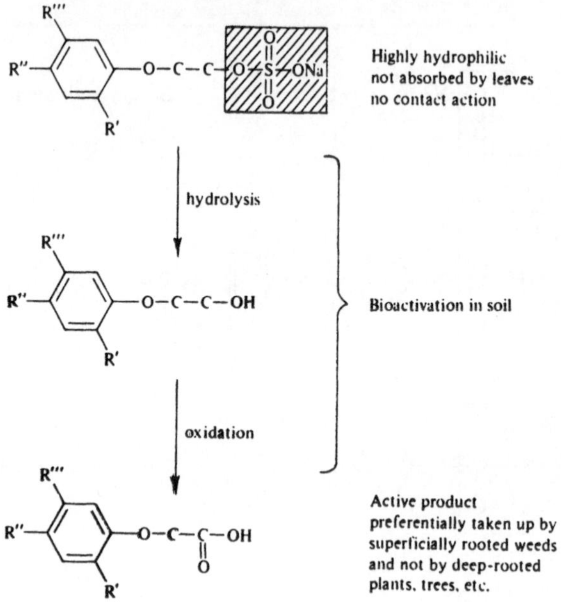

R'	R''	R'''	
—C	—Cl	—H	Methin
—Cl	—Cl	—H	Sesin
—Cl	—Cl	—Cl	Natrin

Fig. 28. Selectivity in action of weed killers obtained by introduction of a highly hydrophilic disposable restricting moïety in a precursor compound. After Crafts[91]

(Fig. 29). Unlike the original radio-contrast media, which, being water-soluble, were easily spread over the bronchial surface and the alveoli and hence rapidly absorbed, the esters, which are water-soluble, adhere to the bronchial wall without penetrating the smaller bronchioli and thus allow good visualization of the bronchial tree. The transport-restricting moieties, the alcohols, are split off gradually by hydrolysis and the original radiopaque is set free. This is then absorbed, so that after a certain period of time the agent is cleared from the bronchial tree.

3.2.4. Introduction of Transport-facilitating Moieties to Enhance Absorption

Molecular manipulation resulting in a lipid/water-solubility suitable for penetration of both lipid membranes and the water compartments such as the interstitial fluids will enhance absorption and facilitate distribution over the various

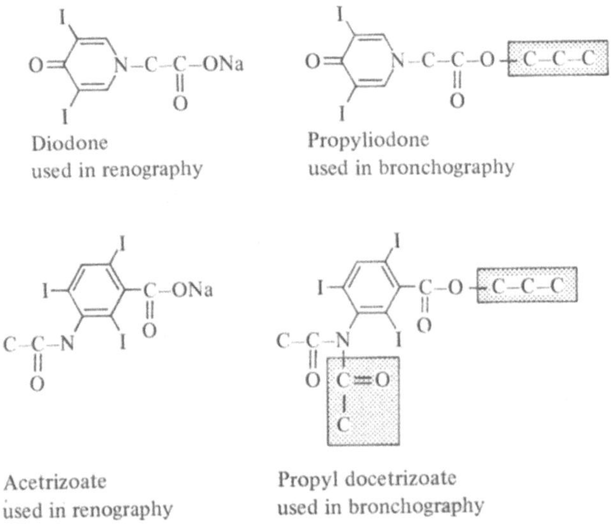

Diodone
used in renography

Propyliodone
used in bronchography

Acetrizoate
used in renography

Propyl docetrizoate
used in bronchography

Fig. 29. Introduction of lipophilizing moieties to restrict the flow of the compound at the site of application

compartments in the organism. The skin forms a real barrier to water-soluble agents but is relatively easily penetrated by lipid-soluble compounds. This is why transport-facilitating moieties are found especially among drugs designed for dermal application. Examples are esters of corticoid hormones, of nicotinic acid, of salicylic acid and the compound clofibrate[147]. The transport-facilitating moieties should be disposable in this case since the original compound (the steroid hormone, nicotinic acid, or salicylic acid) must be set free to exert its action (Fig. 30). Transport-facilitating moieties may also improve the intestinal absorption of orally administered drugs. The penicillins, for instance, are compounds with a free carboxylic acid group and are nearly completely ionized at the high pH of the intestine; they are thus relatively poorly absorbed. Introduction of disposable moieties that mask the ionized carboxylic acid group greatly enhances absorption from the gut. The penicillin esters must resist hydrolysis at the low pH in the stomach and the high pH in the gut but be rapidly hydrolyzed by plasma esterases and possibly esterases in the intestinal wall, so that the active compound is readily set free after absorption (Fig. 31)[148] On the other hand, the penicillin esters should not be too lipophilic since then they become insoluble in water, and instead of a rapid, enhanced absorption, there may be a slow, protracted uptake as with penamecillin[149]. This acetoxymethyl ester of benzylpenicillin is so rapidly hydrolyzed after absorption that only benzylpenicillin can be detected in the blood of the vena portae. The plasma levels obtained with penamecillin persist, however, over a 10 h period,

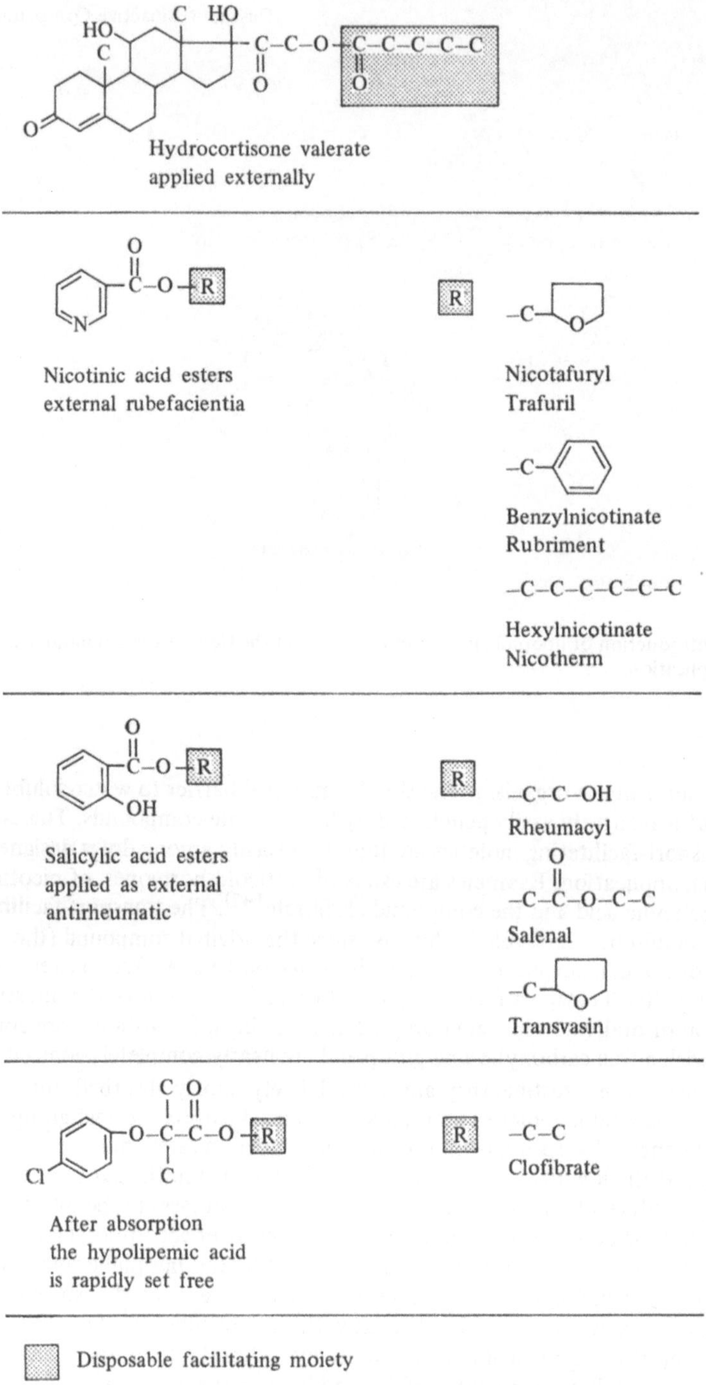

Hydrocortisone valerate
applied externally

Nicotinic acid esters
external rubefacientia

R: Nicotafuryl
Trafuril

Benzylnicotinate
Rubriment

$-C-C-C-C-C-C$

Hexylnicotinate
Nicotherm

Salicylic acid esters
applied as external
antirheumatic

R: $-C-C-OH$
Rheumacyl

$-C-\overset{O}{\overset{\|}{C}}-O-C-C$
Salenal

Transvasin

After absorption
the hypolipemic acid
is rapidly set free

R: $-C-C$
Clofibrate

Disposable facilitating moiety

Fig. 30. Introduction of disposable lipophilizing moieties facilitating absorption

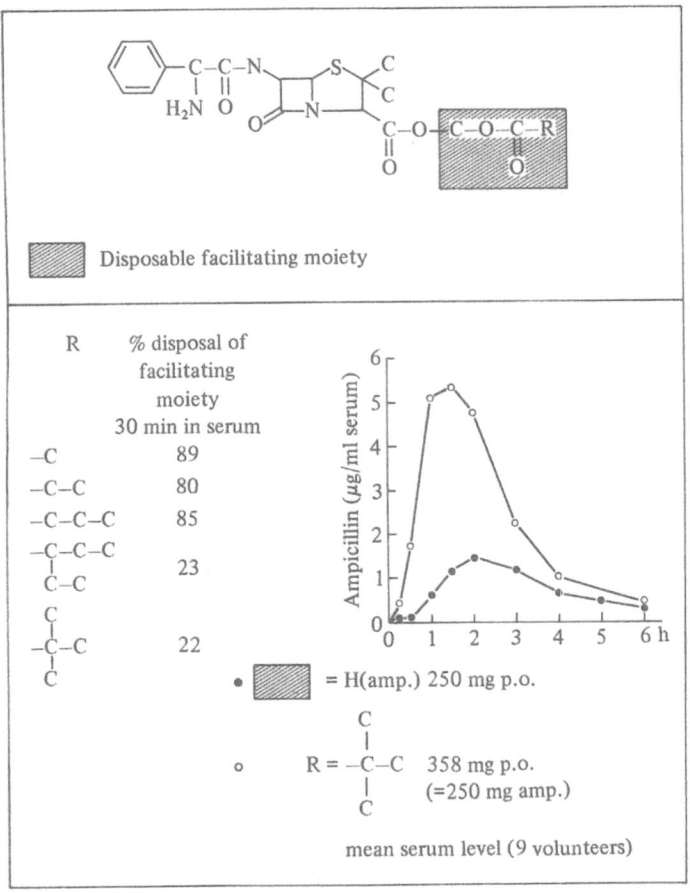

Fig. 31. Introduction of disposable facilitating moieties to enhance intestinal absorption. [From: W. von Daehne et al.[148).]

compared to 5 h for an equivalent dose of benzylpenicillin. With the latter compound, however, much higher plasma levels are reached more rapidly than with penamecillin[149, 157). The fact that ampicillin has an amino group in the side-chain means that after esterification of the carboxyl group a balanced lipid/water-solubility is obtained, allowing for rapid absorption. An analogous approach, namely esterification of one of the carboxyl groups, is used in the conversion of the highly water-soluble carbenicillin, a penicillin with two carboxyl groups, to carindacillin, the indanyl ester of carbenicillin in which one carboxyl group is masked. In this way the very poorly absorbed carbenicillin is converted to a preparation that is active after oral application[150).

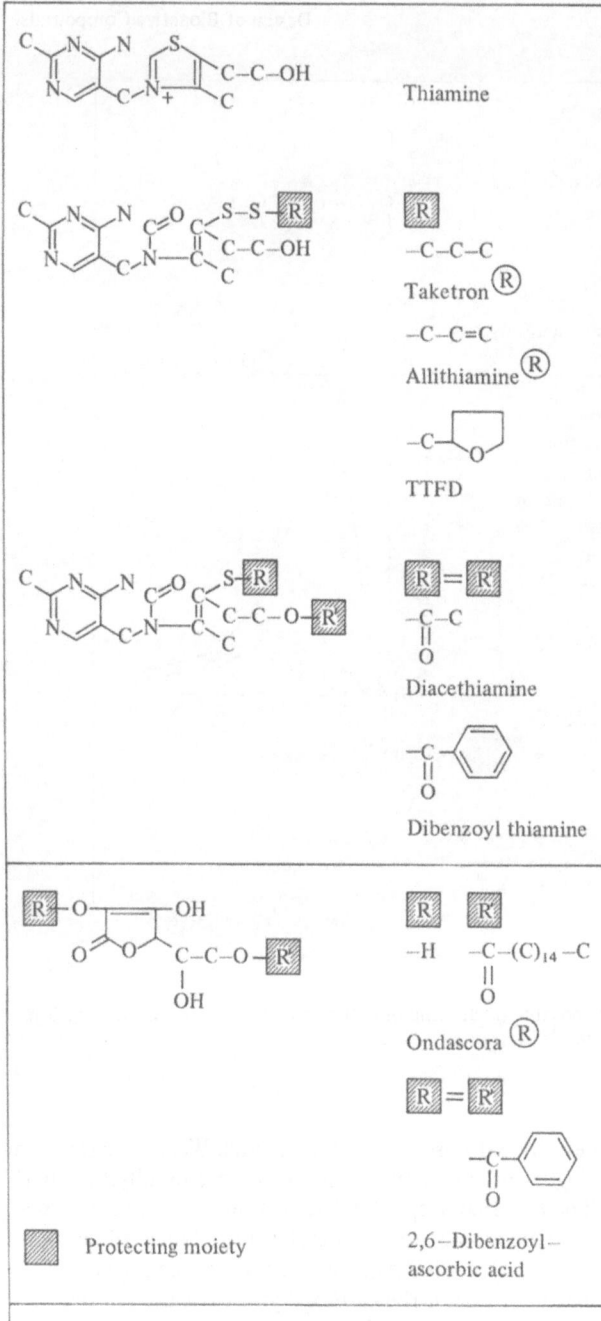

Thiamine

R
−C−C−C
Taketron®
−C−C=C
Allithiamine®

−C
TTFD

R = R
−C−C
‖
O
Diacethiamine

−C
‖
O
Dibenzoyl thiamine

R   R
−H   −C−(C)$_{14}$−C
       ‖
       O
Ondascora®

R = R
−C
‖
O
2,6−Dibenzoyl−
ascorbic acid

▨ Protecting moiety

Less extraction and less decomposition
during food processing and cooking

Fig. 32. Introduction of
disposable facilitating
moieties to enhance
intestinal absorption

Similarly facilitating lipophilizing ester moieties are introduced into highly water-soluble cytostatic antimetabolites, such as methotrexate and derivatives, and into cytarabine in order to enhance penetration of the blood-brain barrier and thus to reach malignant tissue located in the CNS, which is normally out of reach for cytostatics [162, 163].

Disposable transport-facilitating moieties are also used to enhance the absorption of the water-soluble vitamins used as food additives, such as thiamine, ascorbic acid, and riboflavine. The vitamin derivatives obtained are poorly water-soluble and therefore are less extracted during the preparation of the food, which also gives some protection against oxidative decomposition. The increased lipophilicity enhances absorption from the intestinal tract (Fig. 32) [151–155].

The procedure under discussion is not restricted to drugs, as demonstrated in Figure 33, which shows a number of dinitrophenol derivatives used as pesti-cides because of their uncoupling action on the oxidative phosphorylation. In this case, facilitating moieties may be both fixed and disposable. The groups bound to the phenolic OH group must be disposable since a free phenolic OH group is essential for the fungicidal action. The alkyl group in the ring can be

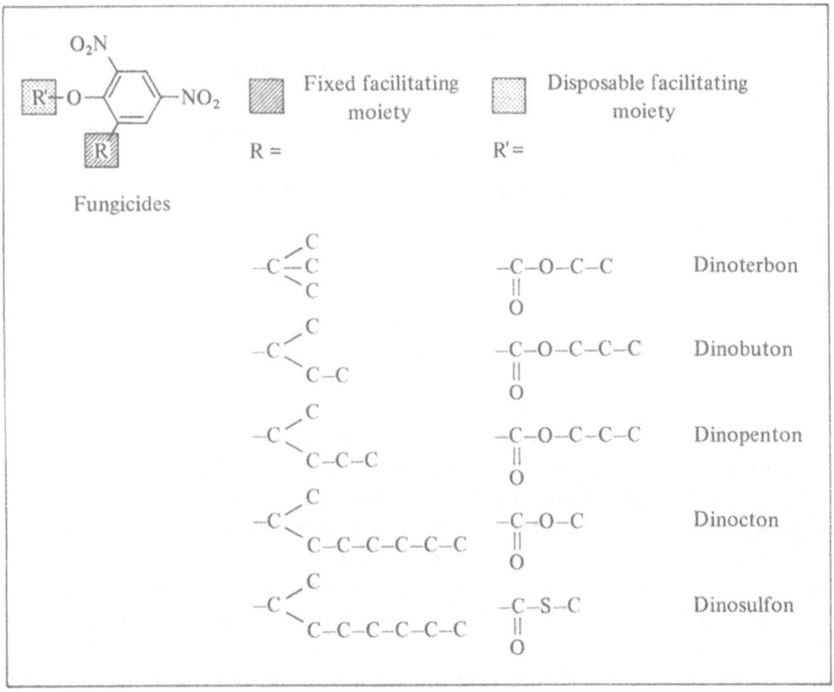

Fig. 33. Introduction of fixed and disposable facilitating moieties in fungicides

regarded as a fixed facilitating moiety. In order to avoid extreme lipophilicity, which would interfere with absorption and transport, the size of the alkyl group on the ring must balance that of the alkyl group on the phenolic OH group[156]. The increased lipophilicity of the derivatives enhances penetration into the fungi, which grow on the surfaces treated and thus come into close contact with the fungicide. The high lipophilicity of the fungicide also prevents its leaching away by water from the surfaces treated. Thus the introduction of the lipophilizing moieties also results in a certain depot action.

3.3. Design of Depot Preparations and Preparations Suitable for Particular Applications

In efforts to modulate the biological availability profile by molecular manipulation, special attention has been paid to preparations with a sustained action, or depot preparations. The use of such depot preparations is most frequent in hormonal substitution therapy. However, depot preparations have also been developed for other types of drugs, usually on basis of conversion of the mother compounds by molecular manipulation to products with low water-solubility. In this respect the slow-release pharmaceutical preparations also deserve a mention. Particular release forms of drugs have been developed to avoid strong local actions and side-effects at the site of application. This is done by introducing disposable moieties that temporarily mask the action of the drug. Finally the hydrophilization of poorly water-soluble drugs for the design of compounds suitable for intravenous application must be mentioned. The pharmaceutical solubilization techniques, as well as the development of suitable sustained-release forms, are aspects of drug design related to the pharmaceutical phase of drug action.

3.3.1. Introduction of Strongly Lipophilizing Disposable Moieties to Obtain Depot Preparations

The principle of obtaining drugs with prolonged action by converting bioactive compounds to water-insoluble products through the introduction of lipophilizing moieties has been applied in various fields. The large families of esters of various steroid hormones and steroid hormone derivatives speak for themselves (Fig. 34). For substitution therapy with hormones a steady, day-long supply of the hormone is required, therefore depot preparations of hormones are frequently used. The various ester preparations are in the form of crystal suspensions or solutions in oil given by intramuscular injection. A particularly interesting ester is the stearoyl glycollate of prednisolone. This compound is given orally; it is rapidly absorbed in the intestinal tract but slowly hydrolyzed by the body esterases and probably also stored in the fat tissues, so giving a long-lasting action. Changes also occur in the spectrum of activities; compared

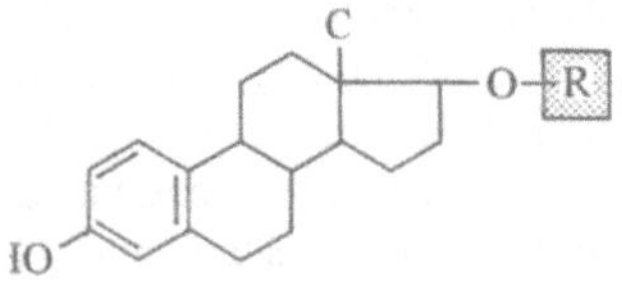

Estradiol   estrogen

R	Propionate
	Valerate
	Heptanoate
	Undecanoate
	Palmitate
	Stearate
	Butyrylacetate
	Trimethylacetate
	Cyclohexanecarboxylate
	β–Cyclopentylpropionate

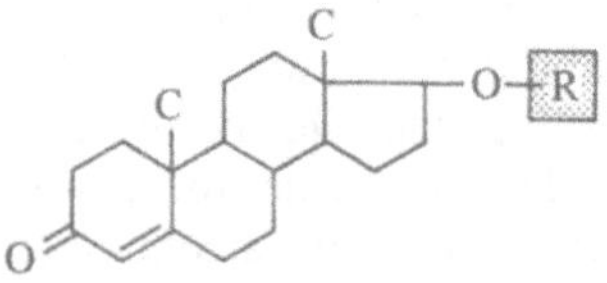

Testosterone   androgen

R	Propionate
	Isobutyrate
	Heptanoate
	Caprinoylacetate
	Cyclohexanecarboxylate
	β–Cyclopentylpropionate
	β–Cyclohexylpropinate
	Phenylacetate
	β–Phenylpropionate
	Nicotinate

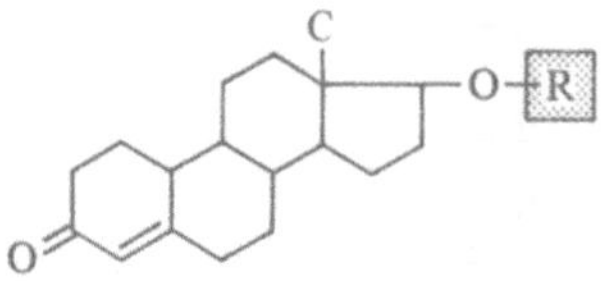

Nandrolone   anabolic

R	Propionate
	Capronate
	Decanoate
	Undecanoate
	Laurate
	Cyclohexanecarboxylate
	β–Cyclopropylpropionate
	β–Cyclohexylpropionate
	Adamantane–3–carboxylate
	β–Phenylpropionate

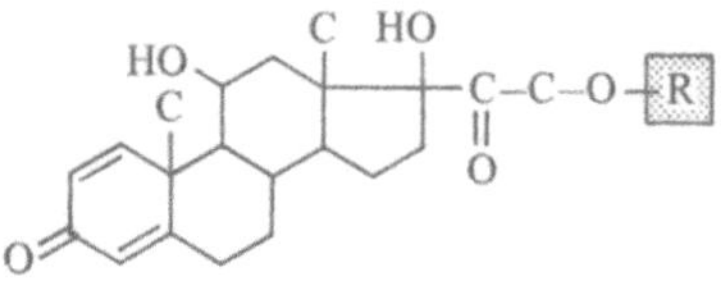

Prednisolone   glucocorticoid

R	Acetate
	Trimethylacetate
	β,β–Dimethylbutyrate
	Stearoyloxyacetate

Fig. 34. Introduction of strongly lipophilizing moieties to obtain depot preparations

to prednisolone, the stearoyl glycollate is more effective as an anti-inflammatory steroid but less active in ulcerogenesis and adrenal suppression[149].

E. J. Ariëns and A.-M. Simonis

The principle of making depot preparations by esterification is widely applied not only to steroid hormones, but also to other drugs (*e.g.* vitamin K, the tranquillizer fluphenazine, and the antimalarial amodiaquine (Fig. 35)[158, 159] and in the field of pesticides to the weedkiller dichlorophenoxyacetic acid. The phosphate ester of the alcohol dichlorophenoxyethanol is extremely lipid-soluble; after application to the soil it stays there as a depot for a long time without being washed away by the rain. It gradually releases the alcohol, which is then rapidly converted by oxidation to the active weedkiller dichlorophenoxy-acetic acid[2].

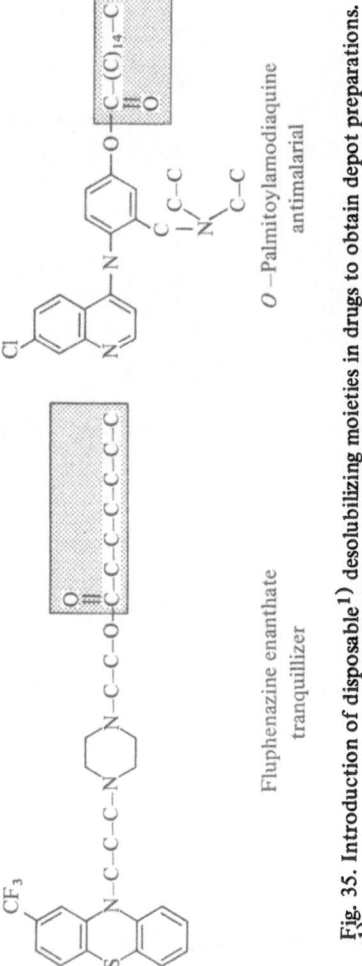

Fluphenazine enanthate
tranquillizer

O −Palmitoylamodiaquine
antimalarial

Fig. 35. Introduction of disposable[1] desolubilizing moieties in drugs to obtain depot preparations.
1) (Disjunction of the desolubilizing moiety before action is highly probable.)

Esterification is also used in the preparation of long-acting fungicides for preserving textiles and paper. The fungicide pentachlorophenol is esterified with lauric acid; the ester is insoluble in water and adheres to the textile fibers so that no leaching occurs if the textile is soaked in water. The active product pentachlorophenol is slowly released and thus guarantees persistent rot-proofing (Fig. 36)[160].

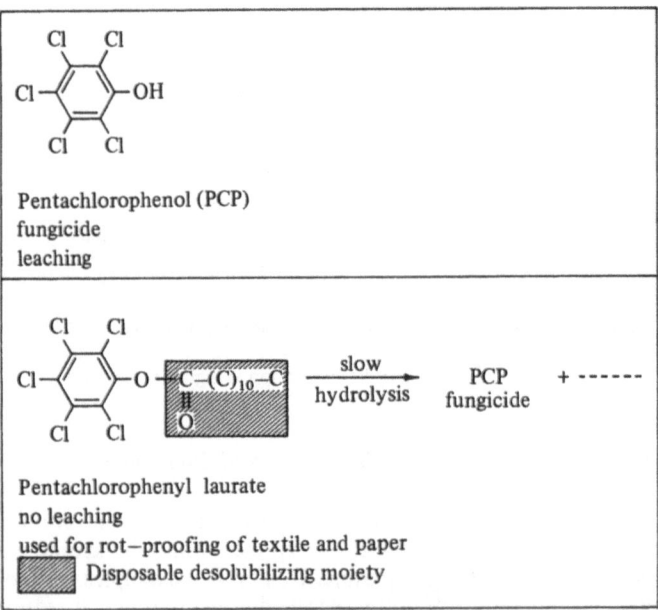

Fig. 36. Introduction of disposable strongly lipophilizing moieties to obtain a depot effect[160]

3.3.2. Introduction of Disposable Hydrophilizing Moieties to Obtain Water-soluble Preparations

A counterpart of the formation of depot preparations on the basis of strongly lipophilizing moieties is the introduction of hydrophilizing moieties into relatively lipophilic compounds. The compounds thus obtained are then suitable for e.g. intravenous application. The hydrophilizing moiety, however, should be easily split off by plasma hydrolases or liver enzymes so that the original compound is set free in its active form. In this way high plasma concentrations can be rapidly built up with poorly water-soluble compounds. The need for disposable hydrophilizing moieties is diminished by the availability of various solubilizing techniques for poorly water-soluble agents, e.g. Tweens. Neverthe-

51

less, the principle is still applied for a number of steroid hormones and other types of drugs such as tranquillizers (Fig. 37).

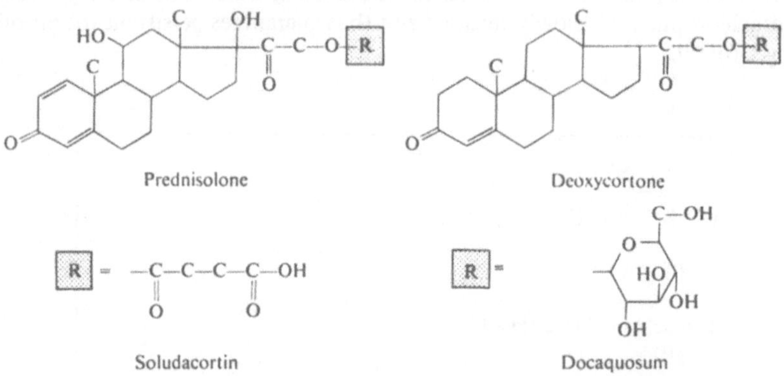

Fig. 37. Introduction of disposable [1]) solubilizing moieties in hormones and hormonoids to obtain water-soluble compounds.
[1]) (Disjunction of the solubilizing moiety before action is highly probable.)

3.3.3. Introduction of Disposable Moieties to Avoid Side-effects

Certain side-effects can be avoided by ensuring that high concentrations of the active compound do not occur at the site of application, or by avoiding initial peak concentrations of the drug in plasma. Well-known examples are the various efforts to mask the phenolic OH group in salicylic acid in order to reduce gastro-intestinal irritation. After absorption, salicylic acid is rapidly set free. The immunosuppressive agent azathioprine is a derivative of the active cytostatic mercaptopurine. After administration of the drug, mercaptopurine is gradually set free, so that high initial peak concentrations do not occur and the risk of a strong cytostatic action is reduced. Masking moieties may also be used to eliminate an unpleasant taste or smell. The extremely bitter taste of chloramphenicol is masked by formation of *e.g.* the palmitate ester, from which the active antibiotic is gradually released in the intestinal lumen (Fig. 38).

4. Conclusion

Modulation of pharmacokinetics in efforts to avoid toxicity or to obtain particular time-effect profiles should be applied in an early phase of drug design. It is effected by means of molecular manipulation based on the elimination or introduction of biofunctional moieties — moieties having a particular biological function, for instance, with regard to biochemical conversion or transport[1, 2, 43, 49−51]. This line of approach has its own specific significance and advantages, and should normally precede the optimization aimed at by the "substituent approach" introduced by Hansch and co-workers[52−56]. In many cases the two approaches are complementary.

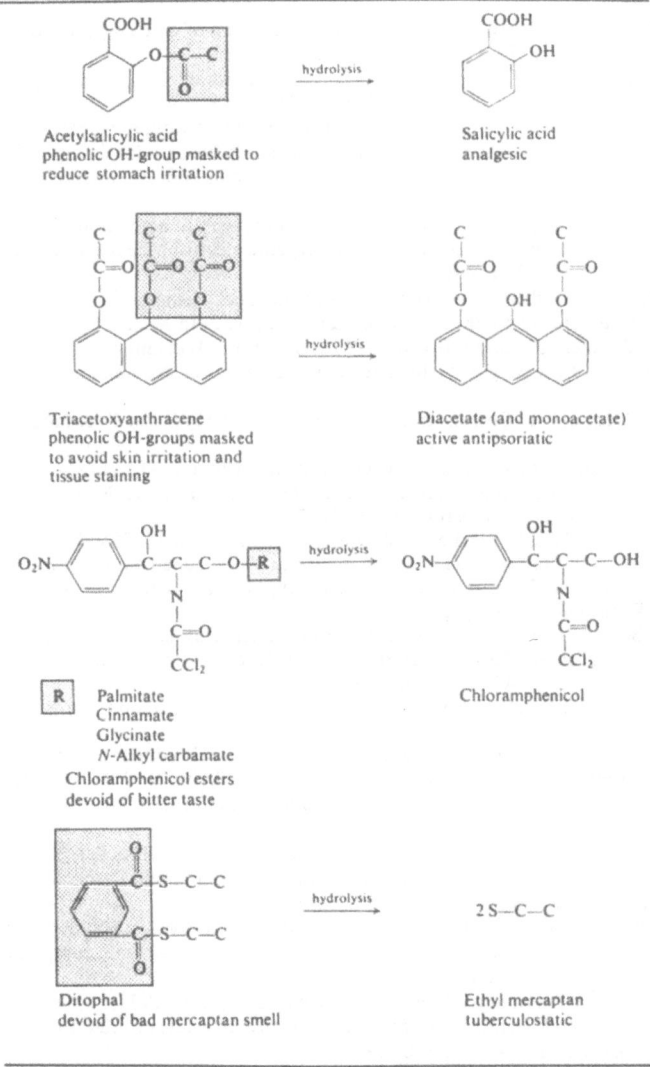

Fig. 38. Introduction of readily disposed masking moieties to obtain compounds with a "latency" in their action and to avoid local "actions" at the site of application

5. References

[1] Ariëns, E. J.: Drug design, Vol. I. New York: Academic Press 1971.
[2] Ariëns, E. J.: Drug design, Vol. II. New York: Academic Press 1971.
[3] Brodie, B. B., Cho, A. K., Krishna, G., Reid, W. D.: Drug metabolism in man: Past, present and future. Ann. N. Y. Acad. Sci. 179, 11 (1971).
[4] Gillette, J. R.: Comparative metabolism and the choice of experimental animals. In: The laboratory animal in drug testing (ed. A. Spiegel). Stuttgart: Gustav Fischer Verlag 1973.
[5] Miller, J. A., Miller, E. C.: The metabolic activation of carcinogenic aromatic amines and amides. Prog. Exp. Tumor Res. 2, 273 (1969).
[6] Miller, J. A., Miller, E. C.: Chemical carcinogenesis: Mechanisms and approaches to its control. J. Natl. Cancer Inst. 47, V (1971).
[7] Miller, J. A., Miller, E. C.: Activation of carcinogenic aromatic amines and amides by N-hydroxylation in vivo. In: Carcinogenesis: A broad critique. Proc. 20th Annual Symp. on Fundamental Cancer Research. Houston 1966. Baltimore: Williams & Wilkins 1967.
[8] Miller, E. C., Miller, J. A.: Studies on the mechanism of activation of aromatic amine and amide carcinogens to ultimate carcinogenic electrophilic reactants. Ann. N. Y. Acad. Sci. 163, 731 (1969).
[9] Miller, E. C., Miller, J. A.: Approaches to the mechanisms and control of chemical carcinogenesis. In: Environment and cancer. Proc. 24th Annual Symp. on Fundamental Cancer Research. Houston 1971. Baltimore: Williams & Wilkins 1972.
[10] Grover, P.: How polycyclic hydrocarbons cause cancer. New Scientist 58, 685 (1973).
[11] Heidelberger, Ch.: Current trends in chemical carcinogenesis. Fed. Proc. 32, 2154 (1973).
[12] Miller, J. A.: Carcinogenesis by chemicals: An Overview – G. H. A. Clowes Memorial Lecture. 60th Annual Meeting of the Am. Ass. for Cancer Research. San Francisco 1969. Cancer Research 30, 559 (1970).
[13] Ryser, H. J. P.: Chemical carcinogenesis. New Engl. J. Med. 285, 721 (1971).
[14] Uehleke, H.: Stoffwechsel von Arzneimitteln als Ursache von Wirkungen, Nebenwirkungen und Toxizität. In: Progress in drug research (ed. E. Jucker), Vol. 15. Basel: Birkhäuser Verlag 1971.
[15] Weisburger, H. H., Weisburger, E. K.: Biochemical Formation and Pharmacological, Toxicological and Pathological Properties of Hydroxylamines and Hydroxamic Acids. Baltimore: Williams & Wilkins 1973.
[16] Peters, R. A.: Biochemical lesions and lethal synthesis. London: Pergamon Press 1963.
[17] Hollaender, A.: Chemical Mutagens, Vol. 1. New York-London: Plenum Press 1971.
[18] Schleiermacher, E., Schroeder, T. M., Adler, I. D., Vrba, M., Vogel, F.: Mutationen durch chemische Einwirkung bei Säuger und Mensch. Dtsch. Med. Wochenschr. 92, 2343 (1967).
[19] Vogel, F., Krüger, J., Röhrborn, G., Schleiermacher, E., Schroeder, T. M.: Mutationen durch chemische Einwirkung bei Säuger und Mensch. Dtsch. Med. Wochenschr. 92, 2382 (1967).
[20] Brodie, B. B., Reid, W. D., Cho, A. K., Sipes, G., Krishna, G., Gillette, J. R.: Possible mechanism of liver necrosis caused by aromatic organic compounds. Proc. Nat. Acad. Sci. 68, 160 (1971).
[21] Clegg, D. J.: Teratology. Ann. Rev. Pharmacol. 11, 409 (1971).
[22] Gillette, J. R., Menard, R. H., Stripp, B.: Active products of fetal drug metabolism. Clin. Pharmacol. Ther. 14, 680 (1973).
[23] Kleiss, E.: Probleme der Teratogenese. Dtsch. Med. Wochenschr. 92, 1507 (1967).
[24] Brodie, B. B.: Enzymatic activation of foreign compounds to more potent or more toxic derivatives. In: Abstracts of Invited Presentations, 5th Int. Congr. Pharmacol. San Francisco 1972.

25) Lewin, R.: Is ageing part of the plan? New Scientist *60*, 615 (1973).

26) Hahn von, H. P.: Primary causes of ageing: a brief review of some modern theories and concepts. Mechanisms of Ageing and Development *2*, 245 (1973).

27) Carr, E. A.: Drug Allergy. Pharmacol. Rev. *6*, 365 (1954).

28) Parker, Ch. W.: The biochemical basis of an allergic drug response. Ann. N. Y. Acad. Sci. *123*, 55 (1965).

29) Thoburn, R., Johnson, J. E., Cluff, L. E.: Studies on the epidemiology of adverse drug reactions. JAMA *198*, 111 (1966).

30) Baer, R. L., Harber, L. C.: Photosensitivity induced by drugs. JAMA *192*, 989 (1965).

31) Jansen, L. H.: Contact sensitivity to simple chemicals: The role of intermediates in the process of sensitization. Naturwissenschaften *51*, 387 (1964).

32) Reid, W. D., Christie, B., Krishna, G., Mitchell, J. R., Moskowitz, J., Brodie, B. B.: Bromobenzene metabolism and hepatic necrosis. Pharmacology *6*, 41 (1971).

33) Mitchell, J. R., Jollow, D. J., Gillette, J. R., Brodie, B. B.: Drug metabolism as a cause of drug toxicity. Drug Metabolism and Disposition *1*, 418 (1973).

34) Reid, W. D., Krishna, G.: Centrolobular hepatic necrosis related to covalent binding of metabolites of halogenated aromatic hydrocarbons. Exp. Mol. Pathol. *18*, 80 (1973).

35) Gillette, J. R., Menard, R. H., Stripp, B.: Active products of fetal drug metabolism. Clin. Pharmacol. Ther. *14*, 680 (1973).

36) Prescott, L. F., Swainson, C. P., Forrest, A. R. W., Newton, R. W., Wright, N., Matthew, H.: Successful Treatment of severe paracetamol overdosage with cysteamine. The Lancet *I*, 588 (1974).

37) Ippen, H.: Mechanisms of photoallergic and phototoxic skin reactions. In: Research progress in organic, biological and medicinal chemistry, Vol. III, Part 2, (eds. U. Gallo and L. Santamaria). Amsterdam: North-Holland Publ. Company 1972.

38) Rodighiero, G., Musajo, L., Dall'Acqua, F., Marciani, S., Caporale, G., Ciavatta, I.: Mechanism of skin photosensitization by furocoumarins. Photoreactivity of various furocoumarins with native DNA and with ribosomal RNA. Biochim. Biophys. Acta *217*, 40 (1970).

39) Wacker, A.: Molecular mechanisms of photodynamic compounds. In: Research progress in organic, biological and medicinal chemistry, Vol. III, Part 1, (eds. U. Gallo and L. Santamaria). Amsterdam: North-Holland Publ. Company 1972.

40) Pathak, M. A.: Photosensitivity to drugs. In: Drugs and enzymes, Vol. 4, (eds. B. B. Brodie, J. R. Gillette). Proc. 2nd Int. Pharmacol. Meeting. Prague 1963.

41) Ariëns, E. J.: Drug levels in the target tissue and effect. In: Clinical pharmacology and therapeutics. Proc. 2nd. Deer Lodge Conf. on Clinical Pharmacology. Hershey (USA) 1973, in press.

42) Ariëns, E. J.: Drug Action: Target tissue, dose-response relationships, receptors. In: Pharmacology and pharmacokinetics. Proc. Conf. Pharmacokinetics: Problems and Perspectives. Washington 1972. New York: Plenum Publ. Company, in press.

43) Ariëns, E. J., Simonis, A. M.: Modulation of the bioavailability profile of drugs (bioactive compounds) by molecular manipulation. In: Proc. FEBS Special Meeting on Industrial Aspects of Biochemistry. Dublin 1973. Amsterdam: North-Holland Publ. Company, in press.

44) Castro, J. A., De Ferreyra, E. C., De Castro, C. R., De Fenos, O. M., Sasame, H., Gillette, J. R.: Prevention of carbon tetrachloride-induced necrosis by inhibitors of drug metabolism — Further studies on their mechanism of action. Biochem. Pharmacol. *23*, 295 (1974).

45) Smolen, V. F., Turrie, B. O., Weigand, W. A.: Drug input optimization: Bioavailability-effected time-optimal control of multiple, simultaneous, pharmacological effects and their interrelationships. J. Pharm. Sci. *61*, 1941 (1972).

46) Schneller, G. H.: Status report on drug bio-availability. Am. J. Hosp. Pharm. *27*, 485 (1970).

47) Wagner, J. G.: Biologic availability, determinant factor of therapeutic activity of drugs. Drug Intelligence and Pharmacy 7, 168 (1973).
48) Dittert, L. W., DiSanto, A. R.: The bioavailability of drug products. J. Am. Pharm. Ass. NS13, 421 (1973).
49) Ariëns, E. J.: Molecular pharmacology, a basis of drug design. In: Progress in drug research, Vol. 10, (ed. E. Jucker). Basel: Birkhäuser Verlag 1966.
50) Ariëns, E. J.: Drug design – Possibilities and limitations. Chimia 26, 355 (1972).
51) Ariëns, E. J.: A molecular approach to the modulation of pharmacokinetics: Modification of metabolic conversion by molecular manipulation. Pure and Applied Chemistry 19, 187 (1969).
52) Hansch, C.: Quantitative structure – Activity relationships in drug design. In: Drug design, Vol. I, (ed. E. J. Ariëns). New York: Academic Press 1971.
53) Hansch, C., Dunn, W. J.: Linear relationships between lipophilic character and biological activity of drugs. J. Pharm. Sci. 61, 1 (1972).
54) Hansch, C., Clayton, J. M.: Lipophilic character and biological activity of drugs II: The parabolic case. J. Pharm. Sci. 62, 1 (1973).
55) Leo, A., Hansch, C., Elkins, D.: Partition coefficients and their uses. Chem. Rev. 71, 525 (1971).
56) Lien, E. J.: Structure-absorption-distribution relationships: Their significance for drug design. In: Drug design, Vol. V, (ed. E. J. Ariëns). New York: Academic Press, in press.
57) Bücher, Th., Sies, H.: Inhibitors: Tools in cell research. Berlin-Heidelberg-New York: Springer 1969.
58) Hochster, R. M., Quastel, J. H.: Metabolic inhibitors, Vols. I–IV. New York: Academic Press 1963–1973.
59) Martin, G. J.: Biological antagonism. Philadelphia: Blakiston 1951.
60) Rhoads, C. P.: Antimetabolites and cancer. London: Bailey Bros. & Swinfen Ltd. 1955.
61) Webb, J. L.: Enzyme and metabolic inhibitors, Vols. I–III. New York: Academic Press 1963, 1966.
62) Wolstenholme, G. E. W., O'Connor, C. M.: Ciba Foundation Symposium on Amino Acids and Peptides with Antimetabolic Activity. London: J. & A. Churchill Ltd. 1958.
63) Wooley, D. W.: A study of antimetabolites. New York: Wiley 1952.
64) Langen, P.: Antimetabolite des Nucleinsäure-Stoffwechsels. Berlin: Akademie-Verlag 1968.
65) Balis, M. E.: Antagonists and nucleic acids. Amsterdam: North-Holland Publ. Company 1968.
66) Varley, A. B.: The generic inequivalence of drugs. JAMA 206, 1745 (1968).
67) Levy, G.: The therapeutic implications of brand interchange. Am. J. Hosp. Pharm. 17, 756 (1960).
68) Ariëns, E. J.: Molecular pharmacology, Vol. I. New York: Academic Press 1964.
69) Baker, B. R.: Design of active-site-directed irreversible enzyme inhibitors. New York: Wiley 1967.
70) Baker, B. R.: Factors in the design of active-site-directed irreversible inhibitors. J. Pharm. Sci. 53, 347 (1964).
71) Baker, B. R.: Specific irreversible enzyme inhibitors. Ann. Rev. Pharmacol. 10, 35 (1970).
72) Baker, B. R.: Metabolite antagonism by enzyme inhibition. In: Medicinal chemistry, Vol. I. (ed. A. Burger). New York: Wiley 1970.
73) Kollonitsch, J., Barash, L., Kahan, F. M., Kropp, H.: New antibacterial agent via photofluorination of a bacterial cell wall constituent. Nature 243, 346 (1973).
74) Post, L. C., Vincent, W. R.: A new insecticide inhibits chitin synthesis. Naturwissenschaften 60, 431 (1973).
75) Editorial: Worms' enzyme differences exploited by chemotherapists. New Scientist 56, 685 (1972).

76) Sanderson, B. E.: Anthelmintics in the study of helminth metabolism. In: Chemotherapeutic agents in the study of Parasites. Symp. Brit. Soc. Parasitology, Vol. *11*, (eds. A. E. R. Taylor and R. Miller) 1973.

77) Hitchings, G. H.: Antimetabolites and Chemotherapy: Integration of biochemistry and molecular manipulation. In: Chemotherapy of cancer. (ed. P. A. Plattner). Amsterdam: Elsevier 1964.

78) Gale, E. F.: Perspectives in chemotherapy. Brit. Med. J. *4*, 33 (1973).

79) Dietschy, J. M. Wilson, J. D.: Regulation of cholesterol metabolism. New Engl. J. Med. *282*, 1128 (1970).

80) Gaunt, R., Steinetz, B. G., Chart, J. J.: Pharmacologic alteration of steroid hormone functions. Clin. Pharmacol. Ther. *9*, 657 (1968).

81) Sobels, F. H.: Chemical mutagenesis and problems involved in the evaluation of environmental mutagens. In: Toxicology: Review and prospect, Vol. XIV. Proc. Eur. Soc. for the Study of Drug Toxicity, Utrecht 1972, (ed. W. A. M. Duncan). Amsterdam: Excerpta Medica 1973.

82) Tuchmann-Duplesis, H.: Drug teratogenicity. In: Toxicology: Review and prospect, Vol. XIV. Proc. Eur. Soc. for the Study of Drug Toxicity, Utrecht 1972, (ed. W. A. M. Duncan). Amsterdam: Excerpta Medica 1973.

83) Loubatières, A.: Utilisation de substances sulfamidées dans le traitement du diabète sucre. Therapie *10*, 907 (1955).

84) Loubatieres, A., Mariani, M. M., Ribes, G., de Malbosc, H., Chapal, J.: Etude expérimentale d'un nouveau sulfamide hypoglycémiant particulièrement actif, le HB 419 ou glibenclamide. I. Action bêtacytotrope et insulino-sécrétrice. Diabetologia *5*, 1 (1969).

85) Loubatières, A., Mariani, M. M.: Pharmacologie. – Étude pharmacologique et pharmacodynamique d'un sulfonylurée hypoglycémiant particulièrement actif, le glybenzyclamide. C. R. Acad. Sci. Paris *265*, 643 (1967).

86) Brodie, B. B.: Of mice, microsomes and man. The Pharmacologist *6*, 12 (1964).

87) Wijngaarden van, I., Soudijn, W.: Difenoxine (R 15 403), the active metabolite of diphenoxylate (R 1132). Arzneim.-Forsch. *22*, 513 (1972).

88) Dvornik, D., Kraml, M., Dubuc, J.: Effect of 22,25-diazacholestanol on synthesis of cholesterol by rat liver homogenates. (29299). Proc. Soc. Exp. Biol. Med. *16*, 537 (1964).

89) O'Brien, R. D.: Effects in plants. In: Toxic phosphorus esters. New York: Academic Press 1960.

90) Wegler, R., Eue, L.: Herbizide. In: Chemie der Pflanzenschutz- und Schädlingsbekämpfungsmittel, Vol. 2, (ed. R. Wegler). Berlin-Heidelberg-New York: Springer 1970.

91) Crafts, A. S.: The chemistry and mode of action of herbicides. In: Advances in pest control research, Vol. I, (ed. R. L. Metcalf). New York: Interscience Publ., Inc. 1957.

92) Röhrborn, G.: Mutations in man and relevant methods for the routine screening of mutagens. In: Toxicology: Review and prospect, Vol. XIV. Proc. Eur. Soc. for the Study of Drug Toxicity, Utrecht 1972, (ed. W. A. M. Duncan). Amsterdam: Excerpta Medica 1973.

93) Editorial: Mutating bacteria spot carcinogenic chemicals. New Scientist *60*, 167 (1973).

94) Newton, B. A.: Trypanocides as biochemical probes. In: Chemotherapeutic agents in the study of parasites. Symp. Brit. Soc. Parasitology, Vol. *11*, (eds. A. E. R. Taylor and R. Muller) 1973.

95) Boyland, E., Williams, K.: An enzyme catalysing a conjugation of epoxides with glutathione. Biochem. J. *94*, 190 (1965).

96) Oesch, F.: Mammalian epoxide hydrases: Inducible enzymes catalysing the inactivation of carcinogenic and cytotoxic metabolites derived from aromatic and olefinic compounds. Xenobiotica *3*, 305 (1973).

97) Evaluation and testing of drugs for mutagenicity: Principles and problems. Wld. Hlth. Org. techn. Rep. Ser. No. 482. Geneva 1971.

98) Elion, G. B.: Biochemistry and pharmacology of purine analogues. Fed. Proc. *26*, 898 (1967).

99) Brodie, B. B., Reid, W. D., Cho, A. K. Sipes, G., Krishna, G., Gillette, J. R.: Possible mechanism of liver necrosis caused by aromatic organic compounds. Proc. Nat. Acad. Sci. *68*, 160 (1971).

100) Editorial: Vitamin C stops carcinogenic chemicals being formed. New Scientist *58*, 70 (1973).

101) Reynolds, E. S., Moslen, M. T.: Liver injury following halothane anesthesia in phenobarbital-pretreated rats. Biochem. Pharmacol. *23*, 189 (1974).

102) Mitchell, J. R., Jollow, D. J., Potter, W. Z., Davis, D. C., Gillette, J. R., Brodie, B. B.: Acetaminophen-induced hepatic necrosis. I. Role of drug metabolism. J. Pharmacol. Exp. Ther. *187*, 185 (1973).

103) Grover, P. L., Sims, P.: K-region epoxides of polycyclic hydrocarbons: Reactions with nucleic acids and polyribonucleotides. Biochem. Pharmacol. *22*, 661 (1973).

104) Ames, B. N., Sims, P., Grover, P. L.: Epoxides of carcinogenic polycyclic hydrocarbons are frameshift mutagens. Science *176*, 47 (1972).

105) Uehleke, H., Hellmer, K. H., Tabarelli-Poplawski, S.: Metabolic activation of halothane and its covalent binding to liver endoplasmic proteins *in vitro*. Naunyn-Schmied. Arch. Pharmacol. *279*, 39 (1973).

106) Levine, B. B.: Immunochemical mechanisms involved in penicillin hypersensitivity in experimental animals and in human beings. Fed. Proc. *24*, 45 (1965).

107) Kalow, W.: Pharmacogenetics. Heredity and the response to drugs. Philadelphia: W. B. Saunders Company 1962.

108) Pharmacogenetics. Wld. Hlth. Org. techn. Rep. Ser. No. 524. Geneva 1973.

109) Evans, D. A. P.: Genetic variations in the acetylation of isoniazid and other drugs. Ann. N. Y. Acad. Sci. *151*, 723 (1968).

110) Vesell, E. S.: Pharmacogenetics. New Engl. J. Med. *287*, 904 (1972).

111) Reynolds, H. T.: Research advances in seed and soil treatment with systemic and nonsystemic insecticides. In: Advances in pest control research, Vol. II, (ed. R. L. Metcalf). New York: Interscience Publ., Inc. 1958.

112) Martin, E. W.: Hazards of medication. Philadelphia: J. B. Lippincott Company 1971.

113) Ariëns, E. J. Simonis, A. M.: The mechanisms of adverse drug reactions. In: Drug-induced diseases, Vol. 4, (eds. L. Meyler and H. M. Peck). Amsterdam: Excerpta Medica 1972.

114) Ariëns, E. J.: Reduction of drug action by drug combination. J. Mondial de Pharmacie *12*, 263 (1969).

115) Ariëns, E. J.: Reduction of drug action by drug combination. In: Progress in drug research, Vol. 14, (ed. E. Jucker). Basel: Birkhäuser Verlag 1970.

116) Beckett, A. H., Rowland, M.: Urinary excretion kinetics of amphetamine in man. J. Pharm. Pharmacol. *17*, 628 (1965).

117) Turner, P., Young, J. H., Paterson, J.: Influence of urinary pH on the excretion of tranylcypromine sulphate. Nature *215*, 881 (1967).

118) Lassen, N. A.: Treatment of severe acute barbiturate poisoning by forced diuresis and alkalinisation of the urine. The Lancet *II*, 338 (1960).

119) Bird, A. E., Marshall, A. C.: Correlation of serum binding of penicillins with partition coefficients. Biochem. Pharmacol. *16*, 2275 (1967).

120) Krieglstein, J.: Zur Plasmaproteinbindung von Arzneimitteln. Klin. Wochenschr. *47*, 1125 (1969).

121) Struller, T.: Progress in sulfonamide research. Prog. Drug Res. *12*, 389 (1968).

122) Cymerman Craig, J., Dwyer, F. P., Glazer, A. N., Horning, E. C.: Tertiary Amine Oxide Rearrangements. I. Mechanism. J. Am. Chem. Soc. *83*, 1871 (1961).

123) Thuillier, M. J., Domenjoz, R.: Zur Pharmakologie der intravenösen Kurznarkose

mit 2-Methoxy-4-allylphenoxyessigsäure-N,N-diäthylamid (G 29505). Anaesthesist 6, 163 (1957).

124) Sigg, E. B., Sigg, T. D., Brinling, J. C.: Autonomic actions of the diethylamide of 2-methoxy-4-allyl-phenoxyacetic acid (G 29505). A Non-Barbiturate Anesthetic. Arch. int. Pharmacodyn. 145, 70 (1963).

125) Thienpont, D., Niemegeers, C. J. E.: Propoxate (R 7464): A new potent anaesthetic agent in cold-blooded vertebrates. Nature 205, 1018 (1965).

126) Janssen, P. A. J., Niemegeers, C. J. E., Schellekens, K. H. I., Lenaerts, F. M.: Etomidate, R-(+)-ethyl-1-(α-methyl-benzyl)imidazole-5-carboxylate (R 16659), a potent, short-acting and relatively atoxic intravenous hypnotic agent in rats. Arzneim.-Forsch. 21, 1234 (1971).

127) Zaagsma, J.: Symposium "Anaesthetic and neuromuscular blocking drugs". Chem. Wkbl. 68, 5 (1972).

128) Pointer, D. J., Wilford, J. B., Bishop, D. C.: Crystal structure of a novel curariform agent. Nature 239, 332 (1972).

129) Bolger, L., Brittain, R. T., Jack, D., Jackson, M. R., Martin, L. E., Mills, J., Poynter, D., Tyers, M. B.: Short-lasting, competitive neuromuscular blocking activity in a series of azobis-arylimidazo-[1,2-α]-pyridinium dihalides. Nature 238, 354 (1972).

130) Huyser, H. W.: Relation between the structure of detergents and their biological degradation. In: Original lectures. 3rd Int. Congr. of Surface Activity. Cologne. Vol. III. 1960.

131) Brooks, G. T.: The design of insecticidal chlorohydrocarbon derivatives. In: Drug design, Vol. IV, (ed. E. J. Ariëns). New York: Academic Press 1973.

132) Holan, G.: Rational design of degradable insecticides. Nature 232, 644 (1971).

133) Beckett, A. H.: In: Enzymes and drug action, pp. 15 and 238, (eds. J. L. Mongar and A. V. S. de Reuck). London: Churchill 1962.

134) Levine, R. M., Clark, B. B.: Relationship between structure and in vitro metabolism of various esters and amides in human serum. J. Pharmacol. exp. Ther. 113, 272 (1955).

135) Thomas, J., Stoker, J. R.: The effect of ortho substitution on the hydrolysis of benzoylcholine. J. Pharm. Pharmacol. 13, 129 (1961).

136) Bird, A. E., Nayler, J. H. C.: Design of penicillins. In: drug design Vol. II, (ed. E. J. Ariëns). New York: Academic Press 1971.

137) Karim, S. M. M., Carter, D. C., Bhana, D., Adaikan Ganesan, P.: Effect of orally administered prostaglandin E₂ and its 15-methyl analogues on gastric secretion. Brit Med. J. 1, 143 (1973).

138) Amy, J. J., Jackson, D. M., Adaikan Ganesan, P., Karim, S. M. M.: Prostaglandin 15 (R) 15-methyl-E₂ methyl ester for suppression of gastric acidity in gravida at term. Brit. Med. J. 4, 208 (1973).

139) Weeks, J. R., DuCharme, D. W., Magee, W. E., Miller, W. L.: The biological activity of the (15S)-15-methyl analogs of prostaglandins E₂ and F₂α. J. Pharmacol. Exp. Ther. 186, 67 (1973).

140) Lippmann, W., Seethaler, K.: Oral anti-ulcer activity of a synthetic prostaglandin analogue (9-Oxoprostanoic acid: AY-22, 469). Experientia 29, 993 (1973).

141) Robert, A., Nylander, B., Andersson, S.: Marked inhibition of gastric secretion by two prostaglandin analogs given orally to man. Life Sci. 14, 553 (1974).

142) Wain, R. L.: The behaviour of herbicides in the plant in relation to selectivity. In: The physiology and biochemistry of herbicides, (ed. L. J. Andus). London: Academic Press 1969.

143) Specifications for identity and purity of food additives. Vol. II: Food colors. Rome: Food and Agriculture Organization of the United Nations 1963.

144) Specifications for Identity and Purity of Food Additives. Volume I: Antimicrobial preservatives and antioxidants. Rome: Food and Agriculture Organization of the United Nations 1962.

145) Hansson, E., Schmiterlöw, G. C.: A comparison of the distribution, excretion and

metabolism of a tertiary (promethazine) and a quaternary (Aprobit®) phenothiazine compound labelled with S 35. Arch. Int. Pharmacodyn. Ther. *131*, 309 (1961).

146) McIsaac, R. J., Koelle, G. B.: Comparison of the effects of inhibition of external, internal and total acetylcholinesterase upon ganglionic transmission. J. Pharmacol. Exp. Ther. *126*, 9 (1959).

147) Havel, R. J., Kane, J. P.: Drugs and lipid metabolism. Ann. Rev. Pharmacol. *13*, 287 (1973).

148) Daehne von, W., Frederiksen, E., Gundersen, E., Lund, F., Mørch, P., Petersen, H. J., Roholt, K., Tybring, L., Godtfredsen, W. O.: Acyloxymethyl esters of ampicillin. J. Med. Chem. *13*, 607 (1970).

149) Launchbury, A. P.: Some recently introduced drugs. In:Progress in medicinal chemistry Vol. 7, (eds. G. P. Ellis and G. B. West). London: Butterworth 1970.

150) Naumann, P., Rosin, H.: Zur oralen Carbenicillin-Therapie mit Carindacillin. Dtsch. Med. Wochenschr. *98*, 2200 (1973).

151) Kawasaki, C.: Modified thiamine compounds. Vitamins and Hormones *21*, 69 (1963).

152) Wagner, H., Wagner-Hering, E.: Lipophilic thiamine derivatives. Med. Klin. *62*, 217 (1967).

153) Editorial: Garlic odor removed from new thiamine derivatives. Japan Medical Gazette November 1964.

154) Mitoma, C.: Metabolic disposition of thiamine tetrahydrofurfuryl disulfide in dog and man. Drug metabolism and Disposition *1*, 698 (1973).

155) Imai, Y.: The antiscorbutic activity of some O-benzoyl derivatives of L-ascorbic acid in guinea pigs. Chem. Pharm. Bull. *14*, 1045 (1966).

156) Schlör, H.: II. Spezieller Teil: Chemie der Fungizide. In: Chemie der Pflanzenschutz-und Schädlingsbekämpfungsmittel, Vol. 2, (ed. R. Wcgler). Berlin-Heidelberg-New York: Springer 1970.

157) Richardson, A., Walker, H. A., Miller, I., Hansen, R.: Metabolism of methyl and benzyl esters of penicillin by different species. Proc. Exp. Biol. Med. *60*, 272 (1945).

158) Kingstone, E.: The use of fluphenazine enanthate in an out-patient clinic. Int. J. Clin. Pharmacol. *1*, 413 (1968).

159) Hirsch, S. R., Gaind, R., Rohde, P. D., Stevens, B. C., Wing, J. K.: Outpatient maintenance of chronic schizophrenic patients with long-acting fluphenazine: Double-blind Placebo Trial. Brit. Med. J. *1*, 633 (1973).

160) Adema, D. M. M., Meijer, G. M., Hueck, H. J.: The biological activity of penta-chlorophenol esters – A preliminary note. Int. Biodetn. Bull. *3*, 29 (1967).

161) Weck de, A. L.: Immunochemical mechanisms in drug allergy. In: Mechanisms in drug allergy, (eds. C. H. Dash and H. E. H. Jones). Edinburgh-London: Churchill Livingstone 1972.

162) Rosowsky, A.: Methotrexate analogs. 2. A facile method of preparation of lipophilic derivatives of methotrexate and 3',5'-dichloromethotrexate by direct esterification. J. Med. Chem. *16*, 1190 (1973).

163) Johns, D. G., Farquhar, D., Chabner, B. A., Wolpert, M. K., Adamson, R. H.: Anti-neoplastic activity of lipid-soluble dialkyl esters of methotrexate. Experientia *29*, 1104 (1973).

164) Sharpe, A.: The physical properties of the carbon-fluorine bond. In: Carbon-fluorine compounds. A Ciba Foundation Symp. Amsterdam: Elsevier-Excerpta Medica 1972.

165) Brodie, B. B.: Enzyme activation of drugs and other foreign compounds to derivatives that produce tissue lesions. In: Pharmacology and the future of man, Vol. 2. Toxicological Problems. Proc. 5th Int. Congr. Pharmacology. San Francisco 1972. Basel: S. Karger 1973.

166) Reid, W. D.: Relationship between tissue necrosis and covalent binding of toxic metabolites of halogenated aromatic hydrocarbons. In: Pharmacology and the future of man, Vol. 2. Toxicological Problems. Proc. 5th Int. Congr. Pharmacology. San Francisco 1972. Basel: S. Karger 1973.

167) Freese, E.: Molecular mechanisms of mutations. In: Chemical mutagens. Principles and methods for their detection Vol. 1, (ed. A. Hollaender). New York: Plenum Press 1971.

168) Miller, E. C., Miller, J. A.: The mutagenicity of chemical carcinogens: Correlations, problems, and interpretations. In: Chemical mutagens. Principles and methods for their detection Vol. 1, (ed. A. Hollaender). New York: Plenum Press 1971.

169) Uehleke, H.: Mechanisms of methemoglobin formation by therapeutic and environmental agents. In: Pharmacology and the future of man, Vol. 2. Toxicological problems. Proc. 5th Int. Congr. Pharmacology. San Francisco 1972. Basel: S. Karger 1973.

170) Poirier, L. A., Weisburger, J. H.: N-Hydroxylation and carcinogenesis. In: Pharmacology and the future of man. Vol. 2. Toxicological Problems. Proc. 5th Int. Congr. Pharmacology. San Francisco 1972. Basel: S. Karger 1973.

171) Gillette, J. R.: Factors that affect the covalent binding and toxicity of drugs. In: Pharmacology and the future of man, Vol. 2. Toxicological Problems. Proc. 5th Int. Congr. Pharmacology. San Francisco 1972. Basel: S. Karger 1973.

172) Boyland, E.: The carcinogenic action of oxidation products of aromatic compounds. In: Biochemical aspects of antimetabolites and of drug hydroxylation, Vol. 16. 5th FEBS Symposium. Prague 1968, (ed. D. Shugar). London: Academic Press 1969.

Received Mai 13, 1974

Antimetabolites:
Molecular Design and Mode of Action

Professor Thomas J. Bardos

Medicinal Chemistry, School of Pharmacy, State University of New York at Buffalo, USA

Contents

1. Introduction

A time span of 33 years has elapsed since the original formulation of the anti-metabolite theory by Fildes[1] based upon the discovery by Woods[2] of the biological antagonism between sulfanilamide (*1b*) and *p*-aminobenzoic acid (*1a*), two compounds strikingly similar to each other in chemical structure and molecular dimensions (see illustration). Up to the present day, this theory

1a   *1b*

has inspired an enormous amount of research work in medicinal chemistry. In its various applications and modifications, it has provided the tools for many fundamental studies of the metabolic pathways and of the structures and functions of various enzymes and cellular receptors, and it served as a basis for most of the major "rational" approaches and strategies employed or proposed in the field of chemotherapy. It may be true that, in the latter field, the anti-metabolite theory has until recently been more useful in providing *post facto* insight into the modes of action of fortuitously discovered chemotherapeutic agents than in the *a priori* design of new drugs displaying unique or superior therapeutic effects; therefore, some of the original expectations of many investigators often resulted in disappointments. One must remember, however, that the main significance of the antimetabolite theory, in its original form, lies in its recognition of the structural relationship between metabolic inhibitors and metabolites; its proposition, to design effective drugs *via* structural alterations of essential metabolites, was a much too general idea for immediately successful application. In retrospect, it is not surprising that of the many thousands of potential "antimetabolites" synthesized by medicinal chemists over the years, only relatively few proved to be effective drugs, and that even some of these were found useful for reasons that could not be predicted by their designers. Much more had to be learned about the essential metabolic processes and their differences in the host and parasites, the structural and functional properties of various key enzymes, the mechanisms of enzyme regulation, the processes of drug transport, activation and elimination, and many other factors that may modify the action or selectivity of drugs, in order to be able to realize the predictive potential of the antimetabolite theory for the design of new drugs in a judicious and effective manner. This phase of the application of the antimetabolite concept in chemotherapy has just only began.

In this article, I shall attempt to review the structure-activity relationships and modes of action of various types of antimetabolites in the light of recent results and present day knowledge, and to discuss certain aspects of the new (as well as old) approaches in this field which may be of interest to those of us who are engaged in the effort to design more effective chemotherapeutic agents. Throughout this discussion, emphasis will be placed upon the effects of antimetabolites on the metabolic processes in the cells rather than on isolated enzyme systems. Although the usefulness of the latter in the primary design and study of new types of inhibitors is well recognized and will be referred to in the text, the specific topic of enzyme inhibitors has been dealt with in two recent review articles[3, 4] to which the reader may turn for further information.

2. Classical Antimetabolites

2.1. Definition

Most of the antimetabolites synthesized and studied during the first two decades of intensive research following the publication of Fildes' hypothesis[1] were close structural analogues of various vitamins, hormones, amino acids, purines, pyrimidines, and other small molecules considered to be "essential metabolites" of living organisms[5, 6]. The term "classical antimetabolites" was introduced[7] to denote "enzyme inhibitors having only a small change in structure" as compared to the natural substrate of a given enzyme. Since many antimetabolites display *in vivo* a variety of biological activities (as it will be seen below), this terminology is used here in a more general sense, to classify all those antimetabolites which differ from the corresponding natural metabolites by replacement of only one or two atoms or groups. It should be emphasized that in our usage the term "classical" does not by any means imply that the design of such antimetabolites should be considered as a historical and, perhaps, obsolete approach. To the contrary, this approach is today as valid and viable as any of the other, more recently proposed rational approaches to chemotherapy; in fact, some of the most recent successes in this field have been achieved with classical-type antimetabolites.

2.2. Classical Antimetabolites as Enzyme Inhibitors

The inhibitory activity of an antimetabolite depends on its successful competition with the corresponding metabolite for the enzyme(s) of which the metabolite is the natural substrate or cofactor. Thus, the *first requirement* for the activity of *any type* of antimetabolite is, that it should be able to combine with the enzyme in such manner as to prevent effective complexing of the enzyme with the normal metabolite. Once the formation of an enzyme-antimetabolite complex has taken place, there are generally two possibilities:

A. The antimetabolite plays the role of a substrate: If the antimetabolite is capable of undergoing the enzyme-catalyzed reaction with the resulting dissociation of the enzyme-antimetabolite complex into (abnormal) product(s) and the free enzyme, then it may be considered an abnormal substrate, or substitute metabolite. As such, it will competitively interfere with the transformation of the normal metabolite; the extent of such interference depends on the relative affinity of the antimetabolite for the enzyme as well as on the rate of its conversion and subsequent release by the enzyme (*i.e.*, the "turn-over" rate of the enzyme-antimetabolite complex). In the extreme (but important) case when the affinity is very high and the turnover rate very low, such antimetabolites act, in effect, as potent enzyme inhibitors, rather than as substitute metabolites (see B.iii below). In the majority of cases, those classical antimetabolites which are capable of undergoing the enzyme-catalyzed reaction, having affinities and conversion rates comparable to those of the corresponding normal metabolites, exert only a partial and temporary inhibition at those steps of the metabolic pathway in which they themselves are metabolized, and therefore, their effective action as metabolic inhibitors will depend on their inhibition of other targets and on subsequent metabolic events (see Section 2.3.).

However, this does not apply to the special situation when (1) the enzyme is a synthetase which catalyzes the formation of a covalent bond between the metabolite (or antimetabolite) and a *second* substrate, and (2) the second substrate is available only in a limited amount. In this case, the antimetabolite competes with the metabolite not only for the enzyme but also for the second substrate, with which it will combine covalently to form an inert product. Although this enzyme-mediated reaction of the antimetabolite is reversible by the corresponding metabolite in a competitive manner, due to its potentially crucial metabolic effect, (*i.e.*, the elimination of another, limiting metabolite which is required for the same reaction step of the metabolic pathway), this reaction *per se* could be responsible for the over-all inhibitory effect of the antimetabolite. That is, in such particular cases, the metabolic target of the inhibitory action of the antimetabolite may be an enzymic reaction step in which it actually plays the role of a substrate. One might think that this type of situation is a rather special and unusual one, as it may be indeed; however, it so happens that the first descovered and still important class of classical and semi-classical antimetabolites, the sulfonamides, appears to act in this manner, as indicated by the results of a recent study[8] (see Section 3.2.).

B. The antimetabolite blocks the enzyme-catalyzed reaction: If the antimetabolite, after complexing with the enzyme, *cannot* complete the catalytic reaction, then its action is that of a typical enzyme inhibitor. While this is the characteristic mode of action of non-classical antimetabolites (see Section 3.1.), classical antimetabolites will act in this manner if their small structural change relative to the corresponding metabolite is one of the following types:

(i) replacement or modification of a group or atom in the "reacting region" of the metabolite, *i.e.*, the part of the molecule which participates in the enzyme-catalyzed reaction;

(ii) a structural change anywhere in the molecule, that significantly alters either the reactivity or the geometric position (with respect to the catalytic site of the enzyme) of the "reacting region" of the metabolite but not those of its "binding" sites to the enzyme;

(iii) a structural change causing a substantial increase in the strength of "binding" of the molecule to the enzyme (*e.g.*, due to additional ionic or hydrogen bonds), together with a significant decrease or complete loss of its ability to undergo the enzyme-catalyzed reaction; this results in the formation of a "kinetically irreversible" enzyme-antimetabolite complex ("stoichiometric" inhibition);

(iiii) the introduction of a chemically reactive group which causes irreversible binding of the antimetabolite to the enzyme *via* covalent bond formation.

Examples

Examples for case (i) are 5'-deoxythymidine (*2e*) and its recently reported[9, 10] 5'-halogeno and 5'-amino dervatives (*2b–d*); particularly the latter are potent inhibitors of thymidine kinase, competitively reversible by thymidine (*2a*). In these analogues, the 5'-hydroxyl group of thymidine is isosterically replaced, therefore, they cannot undergo phosphorylation by thymidine kinase. Another example is the new quinazoline analogue[11] (*3a*) of folic acid (*4a*) in which the pyrazine ring of the pteridine system is replaced by a benzene ring; the latter cannot be reduced by folate reductase. This compound, and its N^{10}-methyl derivative (*3d*), were found to be effective inhibitors of dihydrofolate reductase and, particularly (*3d*), also showed remarkably potent inhibition of thymidylate synthetase[11]. Additional examples of this type of directly acting classical antimetabolites are included among the structural illustrations in Section 2.7.

The examples for case (ii) are somewhat less clear cut. The inhibition of the enzymatic conversion of orotic acid (*5a*) to orotidylic acid by uracil-6-sulfonic acid (*5b*)[12] may be an example for the effect of the replacement of COOH with SO_3H on the electron density of the "reacting region" of the metabolite (ring-N_1-H), since the analogue (*5b*) is not known to be converted to its ribonucleotide. However, better examples of such electronic effects can be found in some antimetabolites acting at the coenzyme level; (see *e.g.*, compound 7f, Section 2.3.). As to the effect of structural modification on the geometric position of the "reacting region", one of the best examples is the D-isomer of the natural L-tryptophan (*6a*) which competitively inhibits the pancreatic tryptophan-activating enzyme, but unlike other classical antimetabolites of L-tryptophan having the L-configuration, it does not undergo the activation reaction[13]. This is understandable, since the binding of tryptophan to the enzyme primarily involves the indole moiety which is the same for both isomers, whereas the position of the "reacting" group (COOH) depends on the configuration of the α-carbon. Similarly, several 9-(ω-hydroxyalkyl)-6,7-dimethylisoalloxazines (*7b–e*), structural analogues of riboflavin (*7a*), show very strong binding to riboflavin kinase but are not phosphorylated by this enzyme[14], presumably because their ω-hydroxyl groups are not appropriately positioned, due either to the shortness or to the hydrophobic character of their side chains.

5 a   R = COOH
  b   R = SO₃H

6 a

7a R = ribityl   R' = CH₃
7b R = CH₂CH₂OH   R' = CH₃
7c R = (CH₂)₂CH₂OH   R' = CH₃
7d R = (CH₂)₃CH₂OH   R' = CH₃
7e R = (CH₂)₄CH₂OH   R' = CH₃
7f R = ribityl   R' = Cl

For case (iii), the most important and fascinating example of a "stoichiometric" inhibitor is the classical antifolate, aminopterin (*4b*), which, due to the isosteric replacement of a OH with an NH_2 group, is capable of complexing with dihydrofolate reductase about 3,000–10,000 times more strongly[7] than folic acid (*4a*) or the normal substrate, dihydrofolic acid, and, at the same time, shows an almost complete loss of ability to undergo the enzyme catalyzed reaction. Obviously, the change in the 4-position has a profound effect not only on the ionization and hydrogen bonding properties (tautomeric structure) of the substituted pyrimidine ring which participates in the "binding", but, at least indirectly (*i.e.*, in the enzyme-bound form), also on the π-electron system of the pyrazine ring, making this ring less susceptible to reduction by the enzyme. Recently, the quinazoline analogues (*3b* and *c*) of aminopterin and methotrexate have been reported to be equally effective inhibitors of dihydrofolate reductase[11, 15]. The approximately 500-fold increase in the inhibitory activity of *3b* compared to that of *3a* for bacterial dihydrofolate reductase should be indicative of the effect of the 4-amino group solely on the "binding" of the pyrimidine ring to this enzyme, since neither *3a* nor *3b* can be reduced enzymatically due to the isosteric replacement of the pyrazine ring in these analogues.

(iiii) Two well-known examples of covalently bound irreversible inhibitors are azaserine (*8*) and DON (*9*), both classical-type antimetabolites of L-glutamine (*10*). Their main site of action is the 5'-phosphoribosyl-formylglycine-amide: L-glutamine amido-ligase catalyzed step of purine synthesis; they inhibit this enzyme for a short time competitively, but then inactivate it irreversibly[16], due to a neighboring group reaction of the $COCHN_2$ moiety with the SH-group of a cysteine residue in the active site of the enzyme ("endoalkylation")[7]. Similar types of irreversible inhibitors of L-asparagine (*11a*) have been reported recently[17], using either $COCHN_2$, $COCH_2Br$ or $COCH_2Cl$ group as the alkylating functions.

COOH	COOH	COOH	COOH
CH-NH$_2$	CH–NH$_2$	CH–NH$_2$	CH–NH$_2$
CH$_2$	CH$_2$	CH$_2$	CH$_2$
O	CH$_2$	CH$_2$	R
C = O	C = O	C = O	
CHN$_2$	CHN$_2$	NH$_2$	*11a* R = CONH$_2$
			11b R = COCHN$_2$
			11c R = COCH$_2$Br
			11d R = COCH$_2$Cl
8	*9*	*10*	

Analogous replacement of the -CO-NH- portion of the pyrimidine ring of orotic acid (*5a*) with -CO-C(N$_2$)-group led to the synthesis of 3-diazocitrazinic acid (*12*), an antimetabolite of orotic acid in *L. bulgaricus*[18]. This analogue is competitively reversible with orotic acid through a narrow concentration range of *12*, but only if the metabolite and the analogue are simultaneously present in the growth medium.

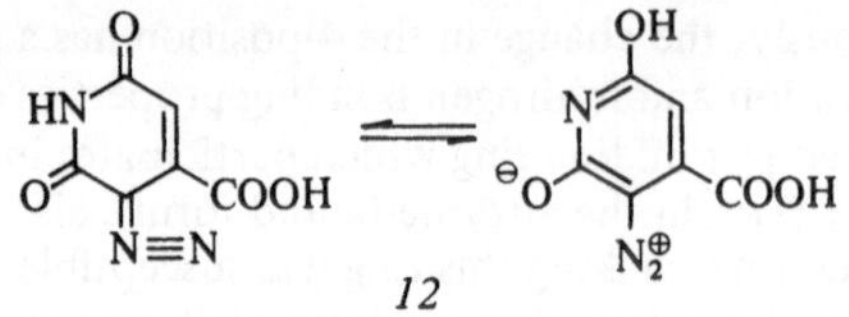

$$12$$

2.3. Lethal Synthesis

Classical-type antimetabolites, due to the close similarity of their over-all structures to those of the corresponding metabolites, can frequently be utilized by the enzymes as substrates and undergo enzymic transformation to yield analogous products. Such antimetabolites may exert their cellular (or viral) inhibitory action *via* one of the following two general mechanisms:

   A. Metabolic activation: The antimetabolites (and their subsequent enzymic reaction product(s), respectively) may be utilized as competitive substrate(s) in one (or several consecutive) enzymic reaction(s) along the metabolic pathway of the normal metabolite, but at one stage of the metabolic reaction sequence, the transformed analogue cannot be further utilized as a substrate and, instead, acts as an inhibitor of the enzyme which catalyzes the next reaction step. At this stage, the action of the *transformed* ("activated") analogue as an enzyme inhibitor depends on the same general types of structural requirements as outlined in the case of the "directly acting" enzyme inhibitors (see Section 2.2.; B (i)-(iiii)).

   B. Incorporation into a functional macromolecule: The antimetabolite may follow the entire reaction sequence of the normal metabolite, ultimately leading to its incorporation into DNA, RNA or protein (enzyme) in which it takes the place of the corresponding normal nucleotide or amino acid unit, respectively. The presence of abnormal nucleotides in the DNA or RNA (1) may inhibit further growth of the polymer chains, (2) may cause genetic mutation (DNA → DNA), or more frequently, (3) erroneous transcription (DNA → RNA) and translation (RNA → protein) of the genetic message; this can result in the synthesis of incomplete or faulty enzymes. The incorporation of amino acid analogues in the place of essential amino acids will also lead to abnormal (sometimes toxic) proteins and non-functioning enzymes. While the effects involving DNA (see (1) and (2), above) may result primarily in the loss of viability of a virus or cell, any of the above listed effects of the incorporation of a "false" precursor into a biopolymer may ultimately cause cell death.

Examples

Examples for mechanism "A" include many vitamin antimetabolites which are metabolically converted to the corresponding coenzyme-analogues; in the latter form, they act as enzyme inhibitors. Thus, 4-deoxypyridoxine (*13b*) and 6-chloro-7-methyl-9-ribityl-isoalloxazine (*7f*), classical antimetabolites of the vitamins B_6 (*13a*) and B_2 (*7a*), respectively, can undergo enzymatic phosphorylation, and then compete with the analogous normal coenzymes for their respective apoenzymes[19]. Inhibition of the subsequent enzyme reaction by the *13b* phosphate would be expected because of the replacement of the catalytic group at the 4-position of the normal coenzyme (*13a* phosphate) by a CH_3 group, while the reported inhibitory effect of the *7f* phosphate[19] may be attributed to the change in the redox-potential of the flavin coenzyme resulting from the isosteric replacement of the 6-CH_3 group with a chlorine atom. (Thus, as enzyme inhibitors, the activated forms of *13b* and *7f* fulfill the criteria B (i) and B (ii), respectively, of Section 2.2.). Antimetabolites of other vitamins acting as enzyme inhibitors only after metabolic conversion, include the following analogues of nicotinic acid (*14a*) and nicotinamide (*14b*): 5-fluoronicotinic acid (*15*), 3-acetylpyridine (*14c*) and 6-aminonicotinamide (*16*). All three of these antimetabolites are enzymatically converted to the corresponding structural analogues of the nicotinamide adenine dinucleotide (NAD) coenzyme, in which form they act as antagonists of NAD in some, if not all of its functions[19].

HO, X, CH_2OH, H_3C, N

$C-X$, O, N

F, COOH, N

$CONH_2$, H_2N, N

13 a X=CH_2OH or CHO *14 a* X=OH *15* *16*
   b X=CH_3 *b* X=NH_2
                              c X=CH_3

In addition, probably most classical antimetabolites of the natural purines and pyrimidines, as well as those of the corresponding nucleosides, which inhibit various enzymes involved in the biosynthesis of nucleic acids, are believed to act primarily by mechanism "A", that is, after metabolic conversion to the corresponding nucleoside-5'-phosphates, diphosphates or triphosphates. Thus, 6-mercaptopurine[20] (6-MP, *17a*) is converted, by inosinic acid : pyrophosphate phosphoribosyltransferase, to 6-thioinosinic acid (TIMP, *17b*), and the latter acts as an effective inhibitor of several important enzymes that catalyze the interconversion of purine nulceotides[21]. 5-Fluorouracil[22] (FU, *18a*) is converted through a series of consecutive enzymatic reaction steps to 5-fluoro-2'-deoxyuridylate[22] (*18c*) which, at first competitively, then irreversibly[23] inhibits thymidylate synthetase. The corresponding nucleoside, 5-fluoro-2'-deoxyuridine[22] (FUDR, *18b*), is directly converted to *18c* by thymidine kinase ("salvage pathway"). Phosphorylation of 5-mercapto-2'-deoxyuridine[24, 25] (MUDR, *19a*) by thymidine kinase leads to the 5'-phosphate (*19b*) which acts as a potent reversible inhibitor of thymidylate synthetase; it appears that the ionized 5-mercapto group participates in the binding of *19b* to the enzyme[25]. 6-Azauracil, as well as 6-azauridine (*20a*) is metabolically converted to 6-azauridylic acid (*20b*) which inhibits the decarboxylation of orotidylic acid (*21*)[26]. The same reaction, a key step of the *de novo* synthesis

71

of pyrimidine nucleotides, is also inhibited by 5-hydroxyuridylic acid[27] (22b), an intermediate metabolic product of 5-hydroxyuridine (22a). Both nucleotide analogues (20b and 22b) contain an electron donating group (ring-N_6: or 5-0⁻, respectively,) to approximate the participation of the 6-COO⁻group of 21 in the binding of the substrate to orotidylic acid decarboxylase. However, 22b readily undergoes further metabolic conversion to the corresponding triphosphate (22c) which is a competitive inhibitor of UTP in the RNA polymerase reaction. The triphosphate 22c is also a substrate for UDP-glucose pyrophosphorylase and is converted to analogues of UDP-glucose and related intermediates of carbohydrate metabolism[28].

| | | | 17 a | 17 b | 18 a | 18 b | X = F | R = H |
| c | X = F | R = PO_3H_2 |

| | | 19 a | X = SH | R = H |
| b | X = SH | R = PO_3H_2 |

20 a	R = H	21	22 a	R = H
b	R = PO_3H_2		b	R = PO_3H_2
			c	R = $(PO_3H)_2PO_3H_2$

There are a large number of nucleoside analogues which undergo conversion to the triphosphate stage before they act as enzyme inhibitors. As such, they may inhibit, in addition to the polymerases, various enzymes which utilize the corresponding normal nucleoside triphosphate (e.g., ATP) as cofactor. Thus, both 9-β-D-xylofuranosyladenine[29] (23) and 9-β-D-arabinofuranosyladenine[30] (*24*) are converted to the corresponding triphosphates, but the triphosphate of *23* seems to inhibit primarily the formation of PRPP (from ATP) and thus it acts on the first step of the *de novo* synthesis of nucleotides [28], while the triphosphate of *24* acts as an inhibitor of DNA polymerase, *i.e.*, in the last step of DNA synthesis[30]. The corresponding cytosine nucleoside, ara-C (*25*) also appears to inhibit DNA synthesis[30] in a similar manner as *24*.

23　　　　*24*　　　　*25*

Examples for mechanism "B", *i.e.*, incorporation into macromolecules, are also numerous, and only a few are mentioned here. 8-Azaguanine (*26*) is readily converted to the corresponding GTP analogue and then incorporated into RNA, replacing a large portion of the guanine residues[31]. In contrast, the antibiotic cordycepin (3'-deoxyadenosine, *27*), after conversion to its triphosphate, is incorporated only into terminal positions of RNA chains and blocks further elongation of the polynucleotide chain[32].

5-Fluorouracil (*18a*), or the corresponding ribonucleoside (FUR)[22], and 4-thiouridine[33], are taken up into RNA in the place of the uridylate moiety, and thus affect protein synthesis[33]; in contrast, 5-bromouracil[34] or, 5-bromo-2'-deoxyuridine[35] (*28a*), as well as 5-iodo-2'-deoxyuridine[36] (*28b*) and 5-trifluoromethyl-2'-deoxyuridine[22] (*28c*), are incorporated into DNA in the

26　　　　　*27*　　　*28 a*　　X = Br
　　　　　　　　　　　　　b　　X = I
　　　　　　　　　　　　　c　　X = CF$_3$
　　　　　　　　　　　　　d　　X = CH$_3$

place of thymidine (*28d*); these analogues may interfere with the replication of cells or viruses. The antibiotic tubercidin (7-deazaadenosine, *36*; see Section 2.6.) is incorporated into both RNA and DNA and effectively inhibits DNA, RNA and protein synthesis[37]. Among the amino acid analogues which are incorporated by bacterial cells or cell-free systems into proteins in the place of the corresponding normal amino acid, 4-fluorophenylalanine[38] (*29*) and 7-azatryptophan[39] (*30*) are shown here, as two typical examples.

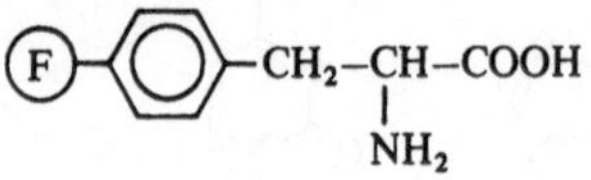

29 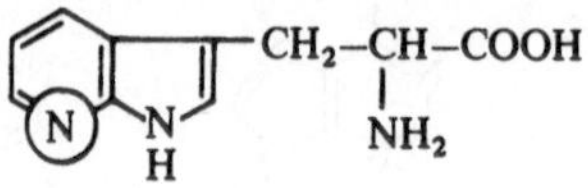   30

2.4. Effects on Regulatory Mechanisms

Classical antimetabolites are often capable of "reinforcing" their effective inhibition of the metabolism and growth of the cell, by acting as "false" feedback inhibitors or repressors, and thus blocking the endogenous biosynthesis of the natural metabolites with which they are to compete for the enzymes as substitute substrates or inhibitors. This type of action depends on a close structural similarity of the antimetabolite to a metabolite which normally regulates its own biosynthesis either by allosteric interaction with an enzyme that catalyzes one of the early reaction steps of the biosynthetic pathway, and thus inhibiting its catalytic function, or by "repressing" the synthesis of the enzymes required for the biosynthetic process. Due to their close structural similarity, such antimetabolites that can act as pseudofeedback inhibitors or repressors, usually also function as substitute metabolites and undergo metabolic activation or incorporation into the macromolecules (see Section 2.3.). However, this is not always the case; *e.g.*, 5-methyltryptophan acts as an enzyme inhibitor (see Section 2.7.).

Some of the previously mentioned pyrimidine nucleoside analogues (FUR, *18b, 22a, 20a, 28a–c*) apparently act in mammalian cells as feedback inhibitors of the *de novo* pyrimidine synthesis at those two early steps of the metabolic pathways which are under regulatory control, *i.e.*, those mediated by the "allosteric" enzymes, aspartate transcarbamylase and dihydroorotate dehydrogenase[40]. The inhibitory activities of these analogues in the isolated enzyme systems are similar to those of the natural nucleosides and of the nucleotide end-products of the *de novo* biosynthesis[40]. In addition, several of these nucleoside antimetabolites, in particular IUDR (*28b*), in the triphosphate form, act as allosteric inhibitors of deoxycytidylate deaminase[41], and of thymidine kinase which is a feedback-controlled enzyme in the "salvage" pathway of DNA-thymine biosynthesis[42].

On the other hand, many of the purine nucleoside analogues act as feedback inhibitors of *de novo* purine synthesis, but only after conversion to their 5'-phosphates. The reaction under regulatory control is the first committed step of purine biosynthesis, the formation of phosphoribosylamine from phosphoribosylpyrophosphate and glutamine. This is normally regulated by AMP and GMP, but some of the analogue nucleotides may be equally or more effective allosteric inhibitors of the transaminase. It has been suggested[43] that the feedback inhibitory effect of its nucleotide may be the most important mechanism of action of 6-MP (*17a*); however, this question is still controversial[21] (see Section 2.6.). Among the other purine nucleoside analogues, the antibiotics cordycepin (*27*) and tubercidin (*36*), in the form of their 5'-phosphates, appear to be effective inhibitors at the regulatory site of *de novo* purine biosynthesis[44].

In bacterial cells, which are able to supply their own amino acid requirements by endogenous synthesis, certain amino acid analogues can act as pseudo-feedback inhibitors or repressors of amino acid biosynthesis, and thereby block the endogenous supply of their "natural competitors"[45]. Prominent examples of pseudofeedback inhibitors are 4-fluorophenylalanine (*29*), 5-methyltryptophan (*56*, see Section 2.7.) and 2-thiazoloalanine (*31a*) which inhibit, at an early step[45], the biosyntheses of phenylalanine, tryptophan, and histidine (*31b*), respectively. Examples of "analogue-repressors" are canavanine (*32*) an analogue of arginine (*33*), trifluoroleucine, and 5-methyltryptophan which shows both feedback inhibitory and repressive properties. (For more details, see Ref.[45].

2.5. Effects on Uptake (Transport) of Metabolites

Amino acids and some other essential metabolites, such as folic acid[46] are taken up by the cells *via* "active transport" mechanisms; that is, they are believed to require specific translocating enzymes ("permeases") to be able to pass through the cell membranes. Close structural analogues of such metabolites may utilize the active-transport mechanism of the corresponding natural metabolite, but apparently less efficiently[45]. Therefore, when an active-transport requiring antimetabolite has to compete with an exogenous supply of the natural metabolite, the latter will gain an initial advantage by reaching a higher concentration than the antimetabolite at least at the first site of their competition for the metabolic enzymes inside the cells[45]. (This can be circumvented by less polar non-classical antimetabolites that can penetrate through the cell membranes *via* passive diffusion, thus by-passing the active-transport mechanism of the metabolite[46]; see Section 3.2.).

However, some classical antimetabolites may derive their chemotherapeutic efficacy from their competitive inhibition of the specific active-transport mechanism of the natural metabolite, provided that the life of the cell depends on a limited exogenous supply of the latter substance. The fascinating "homo-folic acid story"[47] seems to illustrate this point. In brief, "tetrahydrohomo-folate" (*34*) is a potent inhibitor of *E. coli* thymidylate synthetase and was found to prolong the life span of methotrexate-resistant leukemic mice. However, these two activities appear to be unrelated to each other[47]. Based, in part, on studies conducted with various microbiological systems, it was suggested that the *in vivo* activity was due solely to the *d*, L-diastereoisomer of *34* which acts by blocking folate uptake, while the inhibitory activity of the analogue for *E. coli* thymidylate synthetase resides entirely in the *l*, L-diastereoisomer which has the same configuration as the natural cofactor[48].

34

2.6. Multiple Inhibitory Action

Many of the classical-type antimetabolites used as examples in the foregoing section are inherently capable of acting at several different sites of the metabolic pathways and by more than one mechanism, as it was pointed out on various occasions in our discussion. Therefore, it is often difficult to determine which is the most sensitive site of action of a given effective antimetabolite. This question is of importance, however, not only for theoretical reasons, but also from the point of view of obtaining useful "leads" for the design of more effective chemotherapeutic agents. Experimental methods that have been found useful in the exploration of the relative importance of biochemical sites and modes of action of antimetabolites include inhibition analysis[49] and studies using resistant cells lines[43].

Despite continued efforts in many laboratories producing an enormous amount of interesting experimental results and valuable information concerning the complex modes of action of many antimetabolites, the fact remains that even in the longest and best studied cases we are still lacking a completely clear, unambigous, "final" interpretation of the basic mechanisms of the chemotherapeutic action of these drugs. The picture is constantly changing as new results are being reported, even in the case of relatively "simple" drugs that act on a single step of the metabolic pathway (see the already mentioned report on the action of sulfonamides[8]) and much more so in the case of anti-metabolites having multiple sites of action.

For example, it is still not clear to what extent the pharmacologic effects of FU are due to the action of its deoxyribotide as a potent inhibitor of thymi-dylate synthetase, or to the incorporation of its ribonucleotide into the RNA, (or to feedback-inhibition by its nucleosides, or to some other cause), not to speak of the detailed mechanisms of each of these actions which are still sub-jects of intensive investigations and controversies. The same can be said of some still existing differences in results and views concerning the various sites of action[21] of 6-MP.

In the case of some newer agents, *e.g.*, 5-azacytidine[50] (*35*), there are still many discrepancies concerning the results and their interpretation. This potent antimetabolite of cytidine shows many different types of activities, including incorporation into RNA and DNA and powerful inhibitory action on protein synthesis. It has not been possible so far to correlate these effects and their timing[51] and to determine the most significant sites and mechanisms of action of this antimetabolite.

Another interesting example for the multiple sites of action is that of the antibiotic tubercidin[52] (*36*). This 7-deaza analogue of adenosine, a highly active and toxic antineoplastic agent, is readily converted by the appropriate kinases to the triphosphate, and it is incorporated into both RNA and DNA as well as into the NAD coenzyme; in the monophosphate form it also acts a feedback inhibitor of *de novo* purine synthesis. Surprisingly, the most sensitive metabolic target of the primary inhibitory action of this truly versatile anti-metabolite has been identified (for *S. faecalis*) as glycolysis[37]; this was a con-clusion which "could not have been predicted"[52] and which may be of great significance in future efforts to counteract the toxicity of this drug.

35   36

This "multipotentiality" of most classical antimetabolites can be, of course, both an asset and a liability. It can increase the potency as well as the selecti-vity of a drug due to the partial inhibition of several consecutive reaction steps;

on the other hand, it may destroy the selectivity (by increased toxicity) due to some unexpected side effect. One may say that the clinical use of such antimetabolites somewhat resembles a type of combination chemotherapy in which the component drugs are randomly selected and invariably formulated. However, the "selection" and "formulation" are governed by the balanced processes of natural metabolism, and it remains to be seen whether the resulting "combinations" will prove to be more, or less effective in chemotherapy than the "rational" combinations (e.g., dual antagonists) devised by man. Only time will tell.

2.7. Some General Rules of Design

The numerous examples shown so far, correlating the structures of various antimetabolites with their modes of action, should be almost sufficient to illustrate the types of structural modifications that can be employed in the design of classical-type antimetabolites. Therefore, only a brief summary of the "rules" are presented here, with some typical examples. A previous review of the author on this subject has been published[53]. In addition, the design of nucleoside antimetabolites has been recently reviewed comprehensively and critically[52].

A. *Isosteric replacement:* Isosteres, according to Erlenmeyer's definition[54], are "atoms, ions, or molecules in which the peripheral layers of electrons can be considered to be identical". Such isosteres were found to be similar to each other with respect to serological specificity and other types of biological activities. Based essentially on Friedman's classification[55], we may group the biologically important isosteric atoms, "hydrides"[56] and ions, in Classes I–IV, and the isosteric "ring-equivalents"[54], in Class V, as shown in the following table:

Class I	$F, Cl, Br, I; OH, SH, SeH; NH_2; CH_3$
Class II	$O, S, Se, Te; NH; CH_2$
Class III	$N, P, As, Sb, Bi; CH$
Class IV	$C, Si, N^+, P^+, As^+, Sb^+; (S^+)$
Class V	$CH = CH, S (O, NH)$
(in aromatic rings)	

Within each class, the atoms or groups have the same "peripheral layer of electrons" (or, in simpler terms, the same valency), but their dimensions and polarities show considerable variations; nevertheless, replacement of one with the other in biologically active molecules frequently leads to compounds having biological activities either similar or antagonistic to those of the parent compounds.

In Class I, the most effective isosteric replacement is that of a CH_3 group with a Cl, Br, or I atom; F, despite being also an isostere of CH_3, is much smaller in size and, therefore, more effectively substitutes for H. Such replacement is best applicable to the CH_3 substituents of aromatic rings because the resulting aromatic halogens are relatively inert and thus differ less from the CH_3 group in their chemical properties (under physiologic conditions) than the much more reactive alkyl halides. Prominent examples of such isosteric replacement are the 5-bromo and iodo analogues of thymidine (28a, b), the 7-chloro analogue of riboflavin (7f), and 2-chloro-1,4-naphthoquinone, a structural analogue of menadione (vitamin K_3) that has "antivitamin K" activity[57]. Interchange of the other members of Class I also frequently results in antimetabolites. Replacement of OH with NH_2 changes tyrosine (37) into its inhibitory analogue[58], p-aminophenylalanine (38), and guanine into 2,6-diaminopurine[59], an abnormal metabolite undergoing lethal synthesis; however, more dramatic consequences (already discussed in Section 2.2.) of the replacement of OH with NH_2 resulted when it was used to change folic acid (4a) into aminopterin (4b). The reverse replacement of NH_2 with OH changes thiamine into the antivitamin, oxythiamine[60]. Replacement of OH with Cl in hypoxanthine gives 6-chloropurine[31], an effective antimetabolite. Exchange of OH for SH in hypoxanthine leads to 6-MP (17a, Section 2.3. and Section 2.6.) a useful antileukemic drug. Exchange of OH for CH_3 in niacin (14a) and PABA (1a) gives their respective antimetabolites, 3-acetylpyridine (14c) and p-aminoacetophenone[5]; exchange of CH_3 for SH in thymine (39) gives 5-mercaptouracil[61] (MU, 40) a competitive antagonist of thymine in L. leichmannii[62].

37

38

39

40

In Class II, interchange of CH_2 with $-O-$, leads to isosteres with biological activity. Examples are the isosteric pair, azaserine (8) and DON (9) (see Section 2.2.), the amino acids canavanine (32, see Section 2.4.), O-methylthreonine[63] (41) and O-carbamoylserine[38] (43), competitive inhibitors of their respective isosteres, arginine (33), isoleucine (42), and glutamine (10); furthermore, the phosphonate analogues of various nucleotides and other phosphate esters[64, 65] in which replacement of the ester O with CH_2 prevents hydrolysis by the various phosphorolytic enzymes. Similarly, replacement of CH_2 with S in lysine (44) leads to an active competitive analogue, S-(aminoethyl)cysteine[66] (45). Use of Se as replacement for O in hypoxanthine (46) and guanine leads to the effective (but toxic) antimetabolites 6-selenopurine

79

(*47*) and selenoguanine which resemble 6-MP and thioguanine in their anti-tumor properties[67]. The selenium analogue of Coenzyme A has also been synthesized and was found to have partial activity in substituting for Coenzyme A[68].

$$
\begin{array}{ccccc}
\text{CH}_3 & \text{CH}_3 & \text{COOH} & \text{COOH} & \text{COOH} \\
| & | & | & | & | \\
\text{CH}_3-\boxed{\text{O}}-\text{C}-\text{H} & \text{CH}_3-\boxed{\text{CH}_2}-\text{C}-\text{H} & \text{CH}-\text{NH}_2 & \text{CH}-\text{NH}_2 & \text{CH}-\text{NH}_2 \\
| & | & | & | & | \\
\text{H}-\text{C}-\text{NH}_2 & \text{H}-\text{C}-\text{NH}_2 & \text{CH}_2 & \text{CH}_2 & \text{CH}_2 \\
| & | & | & | & | \\
\text{COOH} & \text{COOH} & \boxed{\text{O}} & \boxed{\text{CH}_2} & \boxed{\text{S}} \\
& & | & | & | \\
& & \text{C}=\text{O} & (\text{CH}_2)_2 & (\text{CH}_2)_2 \\
& & | & | & | \\
& & \text{NH}_2 & \text{NH}_2 & \text{NH}_2 \\
\\
41 & 42 & 43 & 44 & 45
\end{array}
$$

In Class III, CH and N are interchanged most frequently in aromatic rings, to give isosteric ring structures such as benzene, pyridine, pyrimidine, pyrazine, and triazine; pyrrole, pyrazole, imidazole, and triazole; thiophen and thiazole; furan and oxazole; as well as the related fused-ring systems. These aromatic rings are also interrelated through isosteric replacement according to the Class V rule; thus, benzene is isosteric with thiophen, pyridine with thiazole, etc. Antimetabolites obtained by this type of isosteric replacement show very close resemblance in biological properties to the corresponding metabolites and may act by any of the mechanisms described in Sections 2.2.–2.6. The examples given there included both aza- and deaza- isosteres of purines, pyrimidines and other heterocycles (*3, 20a, 26, 30, 35, 36*); to these we should add virazole (*48*), the recently discovered broad-spectrum antiviral agent[69]. Shift of a heterocyclic ring-N to other positions in the ring also results in effective anti-metabolites; an important example is the xanthine oxidase inhibitor, allopurinol[70] (*49*), an isostere of hypoxanthine (*46*), used as a drug against gout[71]. Additional examples are the antibiotic formycin[72] (*50*), a C-nucleoside analogue of adenosine with a variety of interesting metabolic activities[52], and 1,2-dehydro-1-(2-deoxy-β-D-*erythro*pentofuranosyl)-2-oxopyrazine 4-oxide (*51*), a potent antimetabolite of deoxyuridine in bacterial systems[73]. Replacement of NH with S was illustrated by the metabolite-antimetabolite pair *31b* and *a* (see Section *2.4.*).

Class V ring-isosteres in which a ring CH=CH group is exchanged with an S atom, are represented by β-2-thienylalanine[38] (*52*), an analogue of phenyl-

$$
\begin{array}{ccc}
46 & 47 & 49
\end{array}
$$

80

alanine (*53*), and the reverse change, by pyrithiamine[5], a long-known anti-metabolite of vitamin B_1 (thiamine).

B. *Replacement of hydrogen:* Exchange of a hydrogen atom with any of the Class I isosteres often leads to antimetabolite activity. Particularly, replacement of H with the spatially similar F results in analogues that are readily utilized by the various enzymes and incorporated into macromolecules (*e.g.*, *18a; 29*), except, when the F substitutes for an H which would undergo displacement in the enzymic reaction (*cp.* action of FUDR-P (*18c*) as inhibitor of thymidylate synthetase; Section 2.3.).

Replacement of H with OH or NH_2 leads to a number of effective anti-metabolites which frequently act as enzyme inhibitors. Thus, β-hydroxyphenyl-alanine[38] (phenylserine; *54b*) is a competitive antagonist of phenylalanine (*54a*) while both β-hydroxyaspartic acid (*55b*) and diaminosuccinic acid (*55c*) are antimetabolites of aspartic acid (*55a*) in bacterial systems[38]. The reverse change of OH to H, often means the elimination of a potential "reacting" group of the substrate (see Section 2.2.; B(i)) and produces enzyme inhibitors[7], or chain terminators (*e.g.*, *27*)[32]. This type of modification may also affect the binding properties (and possibly the positioning) of the molecule at the active site of the enzyme[3].

Replacement of H with CH_3 frequently confers antimetabolite activity. Some of the more interesting examples are the methyltryptophans (*56–58*) (substituted in the 5- or 6-positions of the ring, or in the β-position of the alanine side chain) which inhibit competitively the pancreatic tryptophan-activating enzyme[13]; in addition 5-methyltryptophan acts as both feedback inhibitor and repressor of the endogenous synthesis of tryptophan[45]. Ethionine (*59*), the S-ethyl homolog of methionine (*60*), inhibits protein synthesis, interferes with the methyltransfer reactions of methionine (*via* S-adenosyl-methionine)[73], and exhibits several other antagonistic effects[6].

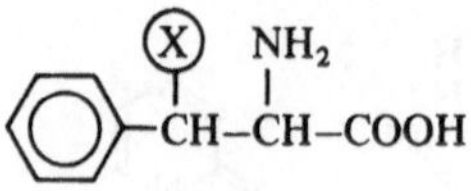

54 a   X = H
     b   X = OH

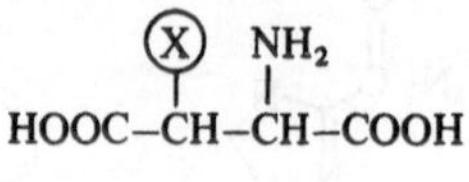

55 a   X = H
     b   X = OH
     c   X = NH$_2$

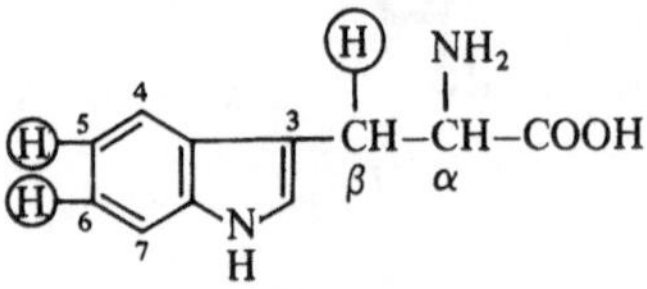

56   5–CH$_3$
57   6–CH$_3$
58   β–CH$_3$

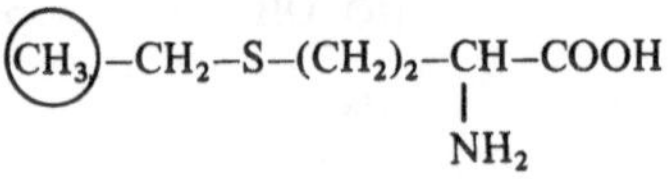

59

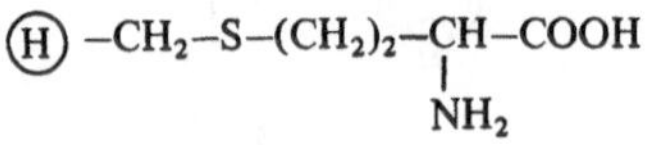

60

C. *Replacement of carboxyl group:* Sulfonic acid analogues of the various amino acids[74] are effective inhibitors of the growth of bacteria, and of the replication of phage and vaccinia virus. Examples of this series are α-amino-methanesulfonic acid, an analog of glycine, and α-aminoisobutanesulfonic acid, an analog of valine[58]. Cysteic acid (*61*) competitively inhibits the conversion of aspartic acid (*55a*), to β-alanine, whereas pantoyltaurin acts as a competitive inhibitor of the enzymatic conversion of pantothenic acid to coenzyme A. Orthanilic acid (*62*) is a reversible inhibitor of the utilization of anthranilate (*63*) in the biosynthesis of tryptophan (*6a*).

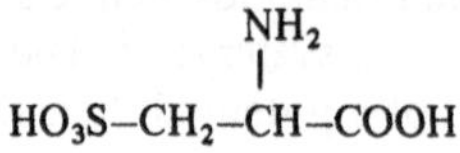

$$HO_3S-CH_2-CH-COOH$$
$$| $$
$$NH_2$$

61

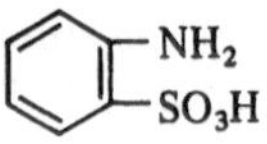

62

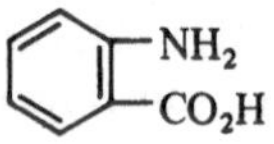

63

3. Non-Classical Antimetabolites

3.1. Definition

The antimetabolites discussed so far are close structural analogues of the corresponding metabolites, having generally similar size and shape, and differing from the latter usually by one atom or chemical group. However, in the design of antimetabolites acting *as enzyme inhibitors per se*, large structural variations are permissible as long as the minimal similarity to the substrate of that part of the molecule which binds to the substrate-specific enzyme site is maintained. Extensive further modifications of the structure of a classical-type antimetabolite have sometimes led to more active or, more selective chemotherapeutic agents. This may be due to favorable changes in the physico-chemical or pharmacologic properties of the compound (such as ionization, solubility, lipid-water partition, absorption, distribution, and metabolic stability), or to the utilization of species differences in some structural details of the target enzyme present in the parasite and host cells. The latter consideration, based on experimental results obtained in the area of antifolates (see Section 3.2. (B)), was first pointed out by Hitchings *et al.*[76] and emphasized again by Baker[77] who postulated that „in order for an antimetabolite to have maximum enzyme specificity, the greatest possible changes in the bulk of the antimetabolite should be made that still allow the stereospecific and binding requirements of the target enzyme to be met". Such enzyme inhibitors, "that have a large but appropriate structural change", Baker termed "non-classical antimetabolites", and, making them the basis of his own systematic and prolific research program aimed specifically at the design of active-site-directed irreversible inhibitors (see Section 3.3.), he developed the theoretical concepts and a rational methodology for this approach[7].

3.2. Non-Classical Antimetabolites Acting as Reversible Enzyme Inhibitors

The activity of a non-classical antimetabolite depends on its ability to complex with the enzyme in such way as to prevent the formation of a functioning enzyme-substrate complex. Due to the large changes in the bulk of the molecule (outside the region which is analogous to the binding site of the substrate), non-classical antimetabolites usually cannot undergo the enzyme-catalyzed reaction; *i.e.*, as a rule, they are not involved in "metabolic activation" (see Section 2.2.); and thus, their only site of action is the original target enzyme.

Theoretically, it is possible to increase both the strength of binding and the enzyme specificity of non-classical antimetabolites acting as reversible inhibitors, by designing the "changed bulk" of the molecule in such way as to utilize certain polar groups or hydrophobic areas of the enzyme, outside the active site, as accessory binding regions. However, since the exact conformational map of the various target enzymes is unknown, the optimal utilization of such accessory binding regions can be established only by trial and error, *i.e.*, by synthesizing and testing a large number of derivatives. A systematic approach to this problem has been developed and demonstrated for several different enzymes[3, 4, 7].

At present, there are two different types of antimetabolites, having large structural changes and acting as reversible enzyme inhibitors, which need to be discussed here. They are represented by two major classes of important chemotherapeutic drugs: the sulfonamides and the antifolates.

T. J. Bardos

A. Semi-Classical Antimetabolites: The Sulfonamides

Soon after the discovery of the antibacterial activity of sulfanilamide (*1b*), empirical structure-activity studies revealed[78] that substitution of the amide nitrogen (N^1) with heterocyclic base radicals leads to much more active antibacterial agents (*e.g.*, sulfathiazole, (*64a*) sulfadiazine, (*64b*) sulfamethoxazole, (*64c*, etc). These derivatives have pK_a values between 6.5 and 8.4 (compared with the pK_a of *1b* at 10.4) and their greater activity was attributed to their more favorable ionization equilibrium at physiologic pH. According to this theory, only the un-ionized molecules penetrate the bacterial cell wall, but only the ionic form competes with the metabolite *1a* (which is ionized) in complexing with the enzyme; thus, sulfonamides that are approximately 50% ionized at physiologic pH would show the most activity. On the other hand, according to an alternative interpretation, the antibacterial activity of the N^1-substituted sulfonamides can be correlated with the negative charge density of the SO_2 group. This is increased by electronattracting substituents on N^1, due to increased ionization, but only up to a point; if the electron-attracting power of the N^1-substituent is too great, this group will compete with the SO_2 group for the electrons surrounding the nitrogen, and the activity decreases[79, 80].

Certainly, the heterocyclic substituents influence not only the ionization but also other physicochemical properties of the drug and may modify the binding of the molecule to the enzyme surface. However, this was not investigated at the time. Recent studies[8] with some of these derivatives (incl. *64c*) indicate that the relatively bulky heterocyclic groups do not interfere with the utilization of the analogue as a substrate in the first step of the dihydrofolate synthetase reaction, that is, the coupling reaction with 7,8-dihydro-6-hydroxymethylpterin. This fact seems to minimize the participation of the bulky group in the complexing of the analogue with the enzyme, and it is for this reason that we tentatively consider this group of agents "semi-classical" antimetabolites.

It has been observed several years ago[81] that cell-free extracts of bacteria were able to incorporate sulfonamides into "folate-like" compounds. That this did not constitute a case of "metabolic activation" (see Section 2.3.), was

84

evident from the fact that synthetic analogues of pteroic acid or folic acid containing a sulfanilamide moiety instead of *p*-aminobenzoic acid (*65a, b*) did not show any biological activity[82]. Therefore, it was difficult to explain the well established facts concerning the complete inhibition of folate synthesis by sulfonamides on the basis of their utilization and conversion by the enzyme to inactive products, and one was tempted to dismiss this phenomenon as a minor "trace" reaction which would not measurably affect the presumed mode of action of these antimetabolites as enzyme inhibitors. However, the results of the recent studies[8] seem to indicate that the inhibitory effect of the sulfonamides might be directly related to their enzymic reaction with the pteridine substrate, resulting in the formation of the inactive "folate-like" products (which have been identified, as *e.g., 65c*) and, consequently, in the elimination of the pteridine, provided that the latter is being produced by the bacteria in a limited amount (see Section 2.2., mechanism A).

65 a   R = H

65 b   R = CH(CO_2H)CH_2CH_2–COOH

65 c   R =

B. Species- and Tissue-Specific Enzyme Inhibitors: The Non-Classical Antifolates

Following the discovery of the classical "antifolate", aminopterin (*4b*), it was found that many other 2,4-diaminopteridines[83] (*e.g., 66*) as well as a number of 2,4-diaminopyrimidine[76] (*e.g., 67a–c*) and 2,4-diamino-1,2-dihydro-*sym*--triazine[84] (*e.g., 68*) derivatives behaved like "folic acid antagonists" in various bacterial systems: their inhibitory effects were reversible with folinic acid rather than folic acid. It was subsequently shown that all these compounds inhibited dihydrofolate reductase. Although they were generally less potent inhibitors than the classical antimetabolites (*4b* and *c*), in contrast to the latter, they exhibited a wide range of selectivities toward different organisms and were also able to discriminate between the dihydrofolate reductase "isozymes" isolated from cells of different species[82]. Of particular interest are the highly selective and potent inhibitory activities of pyrimethamine (*67b*), and trimethoprim (*67c*) against certain bacteria and protozoa; these compounds are important antimalarials. The comparative antibacterial and antiprotozoal activities and the toxicities (in mice) of these drugs and their various congeners showed good correlation with their relative inhibitory activities toward the dihydrofolate reductases isolated from the corresponding cells and tissues. There were large differences noted even between closely related species in the binding of the various inhibitors. From these studies, a "composite map" of the binding regions of these molecules for the dihydrofolate reductases emerged, which identified those binding regions that are common to all and those that are different for the different species[82].

In addition to their species selectivity, another important property of these non-classical antifolates is that, lacking the glutamate moiety, they are capable of crossing through the cell membranes *via* passive diffusion[46] and, thus by-passing the active-transport mechanism on which the cellular uptake of the classical antifolates depends[46]. Recent studies[85] of the pharmacokinetic behavior of *67a* and *b* in animals have shown that these drugs rapidly pass through the blood-brain barrier and can be maintained in the brains at constant and high concentration levels. Particularly *67a* (DMP) showed promise for the clinical treatment of meningeal leukemia which is inaccessible to the classical antifolates.

67 a	$R = CH_3$	$X = 3,4\text{-}di\text{-}Cl$
b	$R = C_2H_5$	$X = 4\text{-}Cl$
c	$R = H$	$X = 3,4,5\text{-}tri\text{-}OH$

In the course of a systematic exploration[7] of the "bulk tolerance" of dihydrofolate reductase, it was discovered that there was a hydrophobic binding region in the area *adjacent* to the active-site, which was responsible for the significant contribution of the non-polar side-chains of the 2,4-diamino-pyrimidine (*67*) and *sym*-dihydrotriazine (*68*) type inhibitors to this enzyme. Consequently, these nonclassical antimetabolites of folic acid are bound at the active site in a "rotomer" position in order to allow their side chains to take an approximately rectangular position with respect to the normal orientation of the glutamate-bearing side chain of folic acid or its classical antimetabolites on the enzyme surface. In exploring further this hydrophobic binding region of the enzyme, a large number of 2,4-diaminopyrimidine and 2,4-diamino-*sym*-dihydrotriazine derivatives were synthesized which contained a variety of long, non-polar side chains linked to the 5-position of the heterocycle. Some of these compounds showed a high degree of selectivity toward various dihydrofolate reductases not only from different species but also from different tissues of the same organism[86, 87]. This tissue-selectivity of these inhibitors apparently resulted from the utilization of fine structural differences of the hydrophobic region of the isozymes from different tissues. Several of these reversible inhibitors showed extremely high potency and selectivity against the dihydrofolate reductase of transplanted tumors, and their *in vivo* antitumor effects correlated with the *in vitro* results. The structure of the "best"[87] reversible antifolate resulting from this work is shown below (*69*).

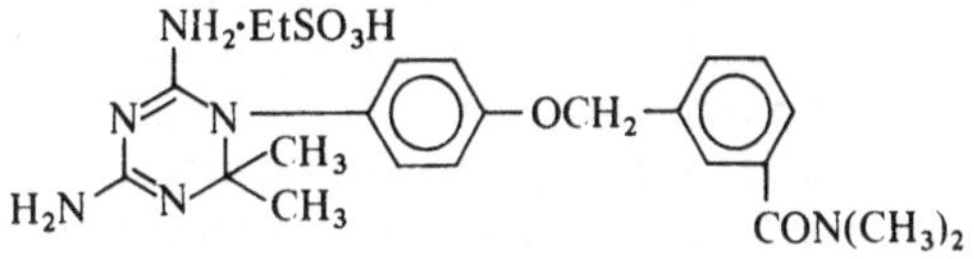

69

Recently, apparently prompted by the promising biological activities[11, 15] of the classical-type quinazoline analogues of folic acid and aminopterin (see Section 2.2., *3a–d*), intensive research programs have been started in several laboratories[88–93] on the synthesis and evaluation of 2,4-diaminoquinazoline derivatives as non-classical antimetabolites of folic acid. During the past two years, an enormous number (probably close to 1000) of different alkyl, aryl and heterocyclic substituted 2,4-diaminoquinazolines have been prepared and reported, – which may reflect, in part, the relative ease and elegance of the general method of synthesis[89, 91] as well as the unusually high ratio of the biologically active compounds obtained in this series, by either sophisticated or routine structural modifications. It seems that, basically, some of these compounds may have been modelled after pyrimethamine and DMP (*67b, a*), as they often retain even the 3,4-dichlorophenyl moiety of the latter at the end of their side chains, in addition to the essential 2,4-diaminopyrimidine nucleus (fused to a benzene ring in the case of the quinazolines). A large number of these derivatives were reported to posses unusually potent, even "prodigious"[92] antibacterial and antimalarial activities (being 200–1160 times more potent than quinine against *Plasmodium berghei* in mice). They act as inhibitors of dihydrofolate reductase but do not exhibit cross-resistance with other antifolates such as pyrimethamine; many of these compounds were highly effective against drug-resistant malaria and Chagas' disease[90]. In addition to the quinazolines, other 2,4-diaminopyrimidino-fused ring systems are also being currently synthesized and evaluated as antifolates and antimalarials[94–96]. At this point, it is not possible to survey this field of rapidly moving research which, in its present phase, seems to be largely empirical. However, in view of its apparent success, this work is bound to advance further our understanding of the modes of action and structure-activity relationships of non-classical antimetabolites, and to produce new leads for the "rational" design of future drugs for chemotherapy. It should be remembered that it was the "rational approach", based on the concepts of antimetabolites and of dihydrofolate reductase inhibitors as chemotherapeutic agents, that has led into and opened up these new, rich fields of potent antiparasitic drugs which are currently being explored.

3.3. Active-Site-Directed Irreversible Enzyme Inhibitors

The concept of active-site-directed irreversible inhibitors has been stated[4] by Baker as follows: "The macromolecular enzyme has functional groups on its surface which logically could be attacked selectively in the tremendously accelerated neighboring group reactions capable of taking place within the reversible complex formed between the enzyme and an inhibitor substituted with a properly placed neighboring group". Thus, the interaction of the inhibitor with the enzyme takes place in two steps: (1) formation of a reversible enzyme-inhibitor complex, and (2) chemical reaction between the reactive (alkylating) group of the inhibitor and a nucleophilic group on the enzyme surface, resulting in the formation of a covalently linked, irreversible enzyme-inhibitor complex. Due to the structural differences between "substrate-identical" enzymes of different species and tissues in regions *outside* their active sites, it is possible to seek out as a target a nucleophilic group of the enzyme surface in a conformational position (relative to the active site) that is specific for the enzyme of the parasite cell, and to design the inhibitor molecule in such a way that its reactive substituent should be placed in juxtaposition with this target when the reversible complex is formed. Such design would provide the irreversible inhibitors with an "extra dimension of specificity" which reversible inhibitors cannot possess; this was termed the "bridge principle of specificity"[7].

Based on this concept, an impressive amount of pioneering work has been carried out during the past decade. The nature of the binding sites and the "bulk tolerances" in the areas adjacent to the active sites of several enzymes have been explored, and some potent and selective inhibitors were synthesized by this systematic approach, which seems to have been carried the farthest and with the most success in the cases of dihydrofolate reductase[86] and adenosine deaminase[3]. For authentic accounts of these highly interesting investigations, a monograph[7] and three, more recent review articles[3, 4, 86] of the original investigators should be consulted.

Finally, it may be worthwhile to point out that one of the most successful chemotherapeutic drugs provided by nature may be considered as a non-classical active-site-directed irreversible inhibitor: This drug is penicillin (*70*). As a non-classical antimetabolite of D-alanyl-D-alanine, it first forms a reversible complex with the enzyme peptidoglycan transpeptidase, and then by a ring-opening reaction of its β-lactam moiety, it forms a covalently linked penicilloyl-enzyme complex[97]. However, the reaction involves the acylation of a sulfhydryl group in the active site of the enzyme; in this respect, *70* resembles the classical-type "endoalkylating" antimetabolites, azaserine and DON (*8* and *9*, see Section 2.2.). Some of the more recently discovered antibiotics and natural products from plants with antitumor activity (*e.g.,* camptothecin) are potential acylating agents, and in view of their highly specific, complex structures, they might be found to provide ideal examples for the "bridge principle of specificity", – once their modes of action become known.

70

4. Dual Antagonists

4.1. Aggregate Analogues

The suggestion that "antimetabolites of unusual and desirable properties may be made by combining in one molecule two analogs of metabolites concerned in closely interrelated and vital functions of living cells" was first made by Woolley[98]. To demonstrate this, he synthesized certain "aggregate analogues" of p-aminobenzoic acid and dimethyldiaminobenzene (the latter was assumed to function as a metabolic precursor of both vitamin B_{12} and riboflavin); theses compounds showed potent inhibitory effects in the *Staphylococcus aureus* test system which, however, could not be reversed either by the structurally related metabolites or by the presumed products of the inhibited reactions. Thus, the actual site(s) and mode(s) of action of these "aggregate analogues" could not be determined.

4.2. Analogues Structurally Related to Two Metabolites

A different type of combination of the structures of two vitamin analogues in one molecule has led to the synthesis of 2,4-diamino-6,7-dimethyl-2,4-deoxyalloxazine (71), the diamino analog of lumichrome[99]. This antimetabolite is structurally related to riboflavin (7a) as well as to the "antifolic" 2,4-diaminopteridines (66). Its potent inhibitory action in *Lactobacillus leichmannii* was, at low concentrations of the inhibitor (0.2-2 µg/ml), competitively reversible with folinic acid alone, but at higher concentrations of LXXI (2–5 µg/ml) both folinic acid and riboflavin were required for reversal. In the *Lactobacillus arabinosus* test system (which does not utilize exogenous folate compounds), thymidine showed non-competitive (product-type) reversal of the inhibition at low concentrations of 71, while thymidine together with riboflavin were effective in overcoming the inhibition at higher concentrations of the inhibitor. Thus, it appears that 71 acts as a competitive, reversible inhibitor for two enzymes, presumably dihydrofolate reductase and riboflavin kinase, but has a much greater affinity for the former; the first enzyme is saturated with the inhibitor, before (at the higher concentrations) the competition of *71* with riboflavin for the second enzyme becomes significant.

71

This approach has the possible advantage over using a combination of two inhibitors, that it eliminates all the pharmacokinetic variables and synchronizes the inhibitory action at the two enzyme sites. However, there is a competition between the two enzymes for the inhibitor, since each molecule of the latter can bind only to one enzyme; therefore, the relative extent of inhibition of the two metabolic reaction steps depends on the relative affinities (ratio of the K_i values) of the inhibitor for the two enzymes. This, of course, is determined by the structure of the inhibitor, and it should be amenable to change *via* structural modifications (*e.g.*, by providing *71* with a hydroxyalkyl side chain and thus making it more closely similar in structure to riboflavin and more antagonistic to it). Although this approach has inherent limitations in scope, its further exploration appears to be of interest.

4.3. "Dual Antagonists" Providing Two Separately Acting Drugs

With the purpose of utilizing the theoretical advantages of combination chemotherapy with optimal effectiveness and selectivity, antimetabolites have been designed in which the biologically essential structural moieties of two different but synergistic antitumor agents are covalently linked to each other, in such manner that they can function separately. Such compounds, termed "dual antagonists", are designed to synchronize the action of the synergistic components and thus overcome differences in absorption, distribution, and elimination that would negate or weaken any potentiation arising from simultaneous block of metabolic pathways at several points. Furthermore, the possibility exists that if two inhibitors, directed against the same biochemical mechanisms or cellular structures but possessing different degrees of selectivity, are coupled chemically, the more selectively localizing component may prevail in directing the distribution of the molecule as a whole. As a result, the less selective (but often more potent) component may reach concentrations at the desired point of attack that could not be attained by the same inhibitor in its free state below its toxic dosage. In particularly favorable cases, both synergistic components may be released at the desired point of attack gradually and simultaneously, providing for a prolonged local action[100, 101].

Based on this rationale, a series of bis(1-aziridinyl)phosphinyl carbamates was synthesized[102] in which "alkylating" ethylenimine derivatives were linked to urethan, an antimetabolite of pyrimidine biosynthesis. Previous studies[103] in animal tumors had indicated that urethan is synergistic with alkylating agents, and it had been strongly suggested in the concluding statements of the New York Academy of Sciences Conferences on Alkylating Agents[104, 105], that combinations of alkylating agents and urethan should receive clinical trial.

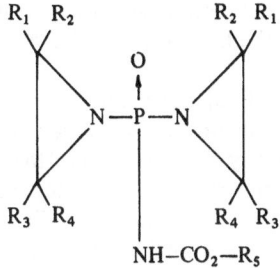

72	$R_1=R_2=R_3=R_4=H$	R_5=ethyl (AB-100)	
73	$R_1=R_2=R_3=R_4=H$	R_5=benzyl (AB-103)	
74	$R_1=R_2=CH_3$	$R_3=R_4=H$	R_5=ethyl (AB-132)
75	$R_1=R_3=CH_3$	$R_2=R_4=H$ (*cis*)	R_5=ethyl (AB-144)
76	$R_1=R_4=CH_3$	$R_2=R_3=H$ (*trans*)	R_5=ethyl (AB-145)

Several members of this "dual antagonist" series showed marked antitumor activities in experimental animals and in man. In both the pharmacologic and subsequent clinical studies, 72 and 73 (in which the aziridine ring carbons are unsubstituted methylene groups) were found to behave essentially like the typical "ethylenimine-type" alkylating agents (*e.g.*, TEPA, or thio-TEPA). Thus, their potent inhibitory activities against a spectrum of transplanted tumors in rodents[106–108], their marked chemotherapeutic activities in various human malignancies[109–112], and their dose-limiting hematologic side-effects, could be attributed to the "alkylating" aziridine function, while the carbamate portion of the molecule appeared to act mainly as a membrane transport "carrier", or as a hydrophobic "binding" group in a more or less selective manner[100]. This was suggested by the fact that the benzyl carbamate analogue 73 appeared to be more effective than the corresponding urethan analogue 72 against several but not all animal tumor systems as well as in the clinical studies against carcinomas of the lung[110] and pancreas[111]. Chemical and NMR studies revealed that during the slow hydrolysis of the aziridine rings of 72 in water at 37 °C (which requires about 8 days for completion) only 13% of the urethan moiety is liberated while most of the urethan remains attached to the phosphorus in the polymerized hydrolysis product[113]. This explains why, at the low dose levels (limited by the toxicity of the C-unsubstituted-aziridinyl function), a direct pharmacologic effect attributable to urethan could not be demonstrated.

Introduction of methyl or ethyl substituents into the aziridine rings generally reduced both the toxicity and the antitumor activity in most of the test systems. The 2,2-dimethylaziridine analogue *74* showed unique chemical as well as pharmacologic properties in this series[114]. Chemical studies indicated that this compound undergoes a very fast ring-opening hydrolysis reaction by a carbonium ion (S_N1) mechanism[100], followed by intramolecular cyclization and subsequent cleavage of all P-N bonds with quantitative liberation of urethan[115]. At the same time, this compound is also capable of participating in alkylation reactions by an S_N2 mechanism at its unsubstituted ring carbon, and it was shown to inhibit the DNA-directed RNA polymerase from *M. lysodeikticus*[116].

Although some significant objective clinical results were obtained with *74* alone, particularly in inoperable bronchogenic carcinoma[117, 118], the remissions obtained were of relatively short duration. However, unusually favorable responses were observed when some patients, after completion of a course of *74* therapy, were treated with X-irradiation[119]. The radiation-potentiating effect of *74* was also observed in animal experiments[120] as well as in the subsequent clinical studies in a large number of cancer patients[121–123].

According to our present view, the radiation-potentiating effect of *74* may be due to the release of urethan in an "activated" (probably, hydroxylated) form. N-hydroxyurethan is known to be similar to N-hydroxyurea in its action as an antimetabolite of DNA biosynthesis. Based on this hypothesis, a new series of dual antagonists was recently synthesized in which the bifunctional alkylating moieties of *72* and *74* were linked to N-hydroxyurethan instead of urethan. Preliminary screening results against animal tumors indicate that these new compounds are much more effective antineoplastic agents than the original series of dual antagonists.

This approach, as compared to the one described in Section 4.2., has a similar pharmacologic rationale, but it is much more flexible in its application and much broader in scope. So far, only a few other types of dual antagonists have been synthesized, including several diaziridines[124] and a series of "arsenical nitrogen mustards"[125], and the results available to date are not sufficient to allow definite conclusions concerning the mode of action of these agents. More work is required in this promising area to confirm the validity of the dual antagonist concept and to explore its further applications in the design of new chemotherapeutic agents.

5. Macromolecular Antimetabolites: Antitemplates

The antimetabolites discussed so far, are all compounds of relatively small molecular weight. In the majority of cases, they act by competitively inhibiting the metabolic transformations of the analogous normal metabolites which usually are intermediates (precursors) or cofactors in the biosyntheses of nucleic acids and proteins. In some other cases, the inhibitory action of the antimetabolites is a consequence of their incorporation into the macromolecules (see Section 2.3. B); however, also this type of action depends on their competition with the normal precursors at the small-molecular (monomer) level. Thus, in order to act selectively, the antimetabolites have to be able to exploit certain biochemical differences between the host and the parasite at the level of intermediary metabolism. If the parasite is a cancer cell, such differences are mostly quantitative rather than qualitative; one would expect, however, that the significant, qualitative differences between the neoplastic and normal cells can be demonstrated at the primary level of genetic expression and information transfer, that is, in their DNA and RNA molecules. It may be possible to exploit the differences between the nucleic acids of the neoplastic and normal cells by preparing structural analogues of the former that would interfere with their functions as templates, messengers or amino acid-transfer RNAs. Such structural analogues of the informational macromolecules may act in an essentially similar manner as the conventional antimetabolites, i.e., by competing with the functional templates for complexing with the DNA and RNA polymerases (particularly, with the "reverse transcriptases") of the cancer cells and forming non-functioning complexes with these enzymes. Therefore, such analogues may be considered as a new class of "macromolecular antimetabolites" and will be termed "antitemplates".

Based on this concept[126], a method has been recently developed which permits the selective introduction of a reactive 5-mercapto substituent into some of the cytosine and/or uracil bases of DNA and RNA molecules without causing degradation or any other chemical alteration[127, 128]. The partially thiolated synthetic and natural polynucleotides were found to be potent inhibitors of DNA-directed RNA polymerase (from *Micrococcus lysodeikticus*) [126]. E.g., a partially thiolated DNA caused 75% inhibition at 10 μg/ml concentration, in the presence of 100 μg/ml (unmodified) DNA template. The inhibition was, in the presence of mercaptoethanol, partially reversible by the (unmodified) DNA template in a concentration-dependent manner, but was irreversible without the addition of mercaptoethanol. The results suggested that the partially thiolated polynucleotides act by competing with the DNA template for the "template site" of the enzyme. In the absence of mercaptoethanol, covalent binding via mixed disulfide linkages may take place between the inhibitor and the enzyme, thus causing irreversible inhibition[128]. On the other hand, DNA polymerase I from regenerating rat liver (which is insensitive to sulfhydryl reagents) was inhibited by thiolated polynucleotides in a reversible manner. In this system, a modified DNA isolate from Ehrlich ascites tumor cells, with 3.9% of its cytosine bases thiolated in the 5-position, showed the highest inhibitory activity; this inhibition was competitively reversed by the corresponding unmodified DNA isolate used as template ($K_i/K_m = 0.5$)[129]. In contrast to both aforementioned polymerases, the DNA polymerase of E. coli-K_{12} was not inhibited by partially thiolated polycytidylic acids[129].

Partially thiolated polycytidylic acid (MPC) and various thiolated DNA and RNA isolates inhibited the DNA polymerases of tumor viruses and showed some significant selectivities[129, 130] (see the article by Dr. Chandra in this volume for more information). Studies of the effects of two, partially thiolated polyribonucleotides on the various types of DNA polymerase activities from cultured human Burkitt lymphoma cells further confirmed the selective inhibitory action of these template analogues[131].

These studies demonstrated that (1) even a relatively minor chemical modification (*e.g.*, introduction of a 5-mercapto group into the cytosine base of 1 out of 100 nucleotide units) can convert a functional DNA or RNA template into a potent, competitive inhibitory analogue (antitemplate) which will bind to the template site of a polymerase either reversibly or irreversibly, and (2) even such antitemplates that are not made to be specific structural analogues of a "natural" template (*e.g.*, MPC) can differentiate between various polymerases. However, there are already some indications that antitemplates more closely related to the "natural" template of a given polymerase are more effective inhibitors of the latter, and it is expected that modified nucleic acids of viruses and tumors will show even much greater selectivities toward the corresponding "reverse transcriptases" in the presence of their endogenous templates.

Some of these antitemplates have shown significant cytotoxicity in bacterial and mammalian cell culture assay systems[126]. There is considerable evidence in the literature[132] that bacterial and mammalian cells are capable of taking up intact DNA and RNA molecules. It has been suggested recently that "one approach to cancer chemotherapy may be the uptake of normal RNA molecules or specifically defined oligonucleotides"[133]. From the point of view of cellular uptake, the increased resistance which these antitemplates demonstrated toward the action of various nucleases appears to be particularly fortuitous.

Different types of antitemplates can be and, presumably, will be, synthesized by employing other chemical reactions that are sufficiently mild and selective for the modification of polynucleotides. Another approach to this problem is, of course, to synthesize an appropriately modified nucleoside triphosphate and utilize it as one of the substrates in a DNA or RNA polymerizing system, in the presence of the specific template for which an antitemplate is to be made. The enzymatic synthesis method has some obvious advantages; however, the inhibitory activity of the product (antitemplate) may limit its applications.

6. Conclusions

The developments in the antimetabolite field during the last few years have been truly exciting. One almost has the feeling that, after the decades of wandering in the desert, we may have, at last, reached the "promised land". Important advances have been achieved on all fronts.

In the area of classical antimetabolites, a host of highly active new nucleosides have been found that show promise in the chemotherapy of cancer and viral diseases. Classical antimetabolites appear to have a virtual monopoly in the field of antiviral agents. After the discovery of the first effective drugs against the *Herpes* virus (*28b, c; 25*), a new breakthrough may have been achieved recently with the appearance of the first broad-spectrum antiviral drug (*48*). The classical antimetabolites have maintained their unquestionable lead also in the field of antineoplastic agents, although they have failed so far to provide really effective and selective drugs for clinical use against cancer. In addition to the nucleoside analogues, some new classical-type antimetabolites of folic acid (*3a–d*, *34*) are of considerable interest *per se* and because they have also provided new leads for the non-classical approach. Allopurinol, although not a chemotherapeutic agent in the sense of Ehrlich, is another important drug that emerged from the classical antimetabolite approach.

The non-classical antimetabolite approach, on the other hand, appears to have struck gold recently in the field of antimalarials as well as other antiprotozoal and antibacterial agents. If the reported animal screening results are any indication of the therapeutic usefulness of the new 2,4-diaminoquinazolines and related compounds, then one would think that the goal of eradicating these parasitic diseases is in sight.

The dual antagonist approach is relatively new, but it has already produced several new anticancer agents which have been showing promise in the clinical trials. This, in itself, is sufficient reason for continuing on with the exploration of this approach, of its further possibilities, or limitations.

The antitemplate approach is, of course, still in the stage of exploratory research. It is certainly an exciting new approach. There is need for new approaches, and hopefully, some more of them will emerge in the near future.

7. References

[1] Fildes, P.: Lancet *1*, 955 (1940).
[2] Woods, D. D.: Brit. J. Exp. Pathol. *21*, 74 (1940).
[3] Schaeffer, H. J., in: Drug design (ed. E. J. Ariëns), Vol. 2, p. 129. New York: Academic Press 1971.
[4] Baker, B. R.: Ann. Rev. Pharmacol. *10*, 35 (1970).
[5] Woolley, D. W.: A study of antimetabolites. New York: J. Wiley and Sons 1952.
[6] Hochster, R. M., Quastel, J. H. (eds.): Metabolic inhibitors, Vol. 1 and 2. New York: Academic Press 1963.
[7] Baker, B. R.: Design of active-site-directed irreversible enzyme inhibitors. New York: J. Wiley and Sons 1967.
[8] Bock, L., Miller, G. H., Schaper, K.-J., Seydel, J. K.: J. Med. Chem. *17*, 23 (1974).
[9] Langen, P., Kowollik, G.: Eur. J. Biochem. *6*, 344 (1968).
[10] Neenan, J. P., Rohde, W.: J. Med. Chem. *16*, 580 (1973).
[11] Bird, O. D., Vaitkus, J. W., Clarke, J.: Mol. Pharmacol. *6*, 573 (1970).
[12] Handschumacher, R. E., Welch, A. D., in: The nucleic acids (eds. E. Chargaff and J. N. Davidson), Vol. 1, p. 453. New York: Academic Press 1960.
[13] Sharon, N., Lipmann, F.: Arch. Biochem. Biophys. *69*, 219 (1957).
[14] Chassy, B. M., Arsenis, C., McCormick, D. B.: J. Biol. Chem. *240*, 1338 (1965).
[15] Hutchison, D. J.: Cancer Chemother. Rept. (Part 1) *52*, 697 (1968).
[16] Levenberg, B., Melnick, I., Buchanan, J. M.: J. Biol. Chem. *225*, 163 (1957).
[17] Chang, P. K., Sciarini, L. J., Handschumacher, R. E.: J. Med. Chem. *16* 1277 (1973).
[18] Papanastassiou, Z. B., McMillan, A., Czebotar, V., Bardos, T. J.: J. Amer. Chem. Soc. *81*, 6056 (1959).

T. J. Bardos

19) Wooley, D. W. in: Metabolic inhibitors (eds. Hochster and Quastel), Vol. 1, pp. 466–474. New York: Academic Press 1963.
20) Hitchings, G. H., Elion, G. B. in: Metabolic inhibitors (eds. Hochster and Quastel), Vol. 1, pp. 215–237. New York: Academic Press 1963.
21) Elion, G. B.: Federation Proceedings, 26 (3), 898 (1967).
22) Heidelberger, C.: Progr. Nucleic Acid Res. Mol. Biol. 4, 1 (1965).
23) Mathews, C. K., Cohen, S. S.: J. Biol. Chem. 238, 367 (1963).
24) Baranski, K., Bardos, T. J., Bloch, A., Kalman, T. I.: Biochem. Pharmacol. 18, 347 (1969).
25) Kalman, T. I., Bardos, T. J.: Mol. Pharmacol. 6, 621 (1970).
26) Handschumacher, R. E., Pasternak, C. A.: Biochim. Biophys. Acta 30, 451 (1958).
27) Smith, D. A., Visser, D. W.: J. Biol. Chem. 240, 446 (1965).
28) Roy-Burman, P.: Analogues of nucleic acid components, pp. 62–64. Berlin – Heidelberg – New York: Springer 1970.
29) Ellis, D. B., LePage, G. A.: Can. J. Biochem. 43, 617 (1965).
30) Cohen, S. S.: Progr. Nucleic Acid Res. Mol. Biol. 5, 1 (1966).
31) Balis, M. E.: Antagonists and nucleic acids. New York: J. Wiley and Sons 1968.
32) Guarino, A. J., in: Antibiotics (eds. D. Gottlieb and P. D. Shaw) Vol. 1, p. 468. Berlin – Heidelberg – New York: Springer 1967.
33) Bloch, A., Kalman, T. I., Bardos, T. J., in preparation.
34) Matthews, R. E. F.: Pharmacol. Rev. 10, 359 (1958).
35) Brockman, R. W., Anderson, E. P.: Ann. Rev. Biochem. 32, 463 (1963).
36) Prusoff, W. H.: Cancer Res. 23, 1246 (1963).
37) Bloch, A., Leonard, R. J., Nichol, C. A.: Biochim. Biophys. Acta 138, 10 (1967).
38) Shive, W., Skinner, C. G.: Ann. Rev. Biochem. 27, 643 (1958).
39) Pardee, A. B., Shore, V. G., Prestidge, L. S.: Biochim. Biophys. Acta 21, 406 (1956).
40) Bresnick, E.: Adv. Enzyme Regulation 2, 213 (1964).
41) Prusoff, W. H., Chang, P. K.: J. Biol. Chem. 243, 223 (1968).
42) Okazaki, R., Kornberg, A.: J. Biol. Chem. 239, 275 (1964).
43) Bennett, L. L., Jr., Simpson, L., Golden, J., Barker, T. L.: Cancer Res. 23, 1574 (1963).
44) Overgaard-Hansen, K.: Biochim. Biophys. Acta 80, 504 (1964).
45) Richmond, M. H.: Biol. Rev. 40, 93 (1965).
46) Wood, R. C., Hitchings, G. H.: J. Biol. Chem. 234, 2377 (1959).
47) Friedkin, M., Crawford, E. J., Plante, L. T.: Ann. N. Y. Acad. Sci. 186, 209 (1971).
48) Kisliuk, R. L., Gaumont, Y.: Proc. 4th Internat. Symp. Pteridines, p. 357. Tokyo: International Academic Printing Co. 1970.
49) Shive, W.: Ann. N. Y. Acad. Sci. 52, 1212 (1950).
50) Jurovčik, M., Raška, K., Jr., Šormová, Z., Šorm, R.: Coll. Czech. Chem. Commun. 30, 3370 (1965).
51) Zain, B. S., Adams, R. L. P., Imrie, R. C.: Cancer Res. 33, 40 (1973).
52) Bloch, A. in: Drug design (ed. E. J. Ariëns), Vol. 4, p. 285. New York: Academic Press 1973.
53) Bardos, T. J.: Hematol. Rev. 3, 53 (1972).
54) Erlenmeyer, H., Leo, M.: Helv. Chim. Acta 15, 1171 (1932).
55) Friedman, H. L.: Natl. Acad. Sci., Natl. Res. Council Publ. No. 206, 295 (1951).
56) Grimm, H. G.: Z. Electrochem. 34, 430 (1928).
57) Woolley, D. W.: Proc. Soc. Exptl. Biol. Med. 60, 225 (1945).
58) Dittmer, K.: Ann. N.Y. Acad. Sci. 52, 1274 (1950).
59) Hitchings, G. H., Elion, G. B., Falco, E. A., Russel, P. B., Vanderwerff, H: Ann. N.Y. Acad. Sci, 52, 1318 (1950).
60) Williams, R. J., Eakin, R. E., Beerstecher, E., Jr., Shive, W.: The biochemistry of B vitamins, p. 697. New York: Reinhold Publ. Co. 1950.
61) Bardos, T. J., Herr, R. R., Enkoji, T.: J. Amer. Chem. Soc. 77, 960 (1955).
62) Bardos, T. J., Levin, G. M., Herr, R. R., Gordon, H. L: J. Amer. Chem. Soc. 77, 4279 (1955).

63) Rabinovitz, M., Olsen, M. E., Greenberg, D. M.: J. Amer. Chem. Soc. 74, 411 (1952).
64) Wolff, M. E., Burger, A.: J. Amer. Pharm. Assoc. 48, 56 (1959).
65) Hullar, T. L.: J. Med. Chem. 12, 58 (1969).
66) Rabinovitz, M., Tuve, R.: Proc. Soc. Exptl. Biol. Med. 100, 222 (1959).
67) Mautner, H. G., Chu, S. H., Jaffe, J. J., Sartorelli, A. C.: J. Med. Chem. 6, 36 (1963).
68) Günther, W. H. H., Mautner, H. G.: J. Amer. Chem. Soc. 87, 2708 (1965).
69) Sidwell, R. W., Huffman, J. H., Khare, G. P., Allen, L. B., Witkowski, J. T., Robins, R. K., Science, 177, 705 (1972).
70) Feigelson, P., Davidson, J. D., Robins, R. K.: J. Biol. Chem. 226, 993 (1957).
71) Rundles, R. W., Elion, G. B., Hitchings, G. H.: Bull Rheum. Dis. 16, 400 (1963).
72) Hori, M., Ito, E., Takita, T., Koyoma, G., Takeuchi, T., Umezawa, H.: J. Antibiotics (Tokyo), Ser. A., 17, 96 (1964).
73) Berkowitz, P. T., Bardos, T. J., Bloch, A.: J. Med. Chem. 16, 183 (1973).
74) Farber, E., Shull, K. H., Villa-Trevino, S., Lombard, B., Thomas, M.: Nature 203, 34 (1964).
75) Shive, W., Skinner, C. G. in: Metabolic inhibitors (eds. R. M. Hochster and J. H. Quastel), Vol. 1, p. 1. New York: Academic Press 1963.
76) Hitchings, G. H., Falco, E. A., Vanderwerff, H., Russell, P. B., Elion, G. B.: J. Biol. Chem. 199, 43 (1952).
77) Baker, B. R., Lee, W. W., Skinner, W. A., Martinez, A. P., Tong, E.: J. Med. Pharm. Chem. 2, 633 (1960).
78) Roblin, R. O., Jr., Williams, J. H., Winnek, P. S., English, J. P.: J. Amer. Chem. Soc. 62, 2002 (1940).
79) Bell, P. H., Roblin, R. O., Jr.: J. Amer. Chem. Soc. 64, 2905 (1942).
80) Roblin, R. O., Jr.: Ann. Rev. Biochem. 23, 501 (1954).
81) Brown, G. M.: J. Biol. Chem. 237, 536 (1962).
82) Hichings, G. H., Burchall, J. J.: Adv. Enzymol. 27, 417 (1965).
83) Daniel, L. J., Norris, L. C., Scott, M. L., Heuser, G. F.: J. Biol. Chem. 169, 689 (1947).
84) Modest, E. J., Foley, G. E., Pechet, M. M., Farber, S.: J. Amer. Chem. Soc. 74, 855 (1952).
85) Stickney, D. R., Simmons, W. S., DeAngelis, R. L., Rundles, R. W., Nichol, C. A.: Proc. Amer. Assoc. Cancer Res. 14, 52 (1973).
86) Baker, B. R.: Accounts Chem. Res. 2, 129 (1969).
87) Baker, B. R.: Ann. N.Y. Acad. Sci. 186, 214 (1971).
88) Davoll, J., Johnson, A. M.: J. Chem. Soc. C, 997 (1970).
89) Rosowski, A., Marini, J. L., Nadel, M. E., Modest, E. J.: J. Med. Chem. 13, 882 (1970).
90) Davoll, J., Johnson, A. M., Davies, H. J., Bird, O. D., Clarke, J., Elslager, E. F.: J. Med. Chem. 15, 812 (1972).
91) Elslager, E. F., Clarke, J., Werbel, L. M., Worth, D. F., Davoll, J.: J. Med. Chem. 15, 827 (1972).
92) Elslager, E. F., Bird, O. D., Clarke, J., Perricone, S. C., Worth, D. F., Davoll, J.: J. Med. Chem. 15, 1138 (1972).
93) Ashton, W., Walker, F. C. III., Hynes, J. B.: J. Med. Chem. 16, 694 (1973).
94) Elslager, E. F., Jacob, P., Werbel, L. M.: J. Het. Chem. 9, 775 (1972).
95) Roskowski, A., Chaykovsky, M., Chen, K. K. N., Lin, M., Modest, E. J.: J. Med. Chem. 16, 185 (1973).
96) Elslager, E. F., Clarke, J., Jacob, P., Werbel, L. M., Willis, J. D.: J. Het. Chem. 9, 1113 (1972).
97) Strominger, J. L., in: Inhibitors tools in cell research (ed. T. Bucher and H. Sies) p. 187. Berlin – Heidelberg – New York: Springer 1969.
98) Woolley, D. W.: J. Am. Chem. Soc. 74, 5450 (1952).
99) Bardos, T. J., Olsen, D. B., Enkoji, T.: J. Amer. Chem. Soc. 79, 4704 (1957).
100) Bardos, T. J.: Biochem. Pharmacol. 11, 256 (1962).

101) Bardos, T. J., Papanastassiou, Z. B., Segalof, A., Ambrus, J. L.: Nature *183*, 399 (1959).
102) Papanastassiou, Z. B., Bardos, T. J.: J. Med. Pharm. Chem. *5*, 1000 (1962).
103) Skipper, H. E.: Cancer *2*, 475 (1949).
104) Gellhorn, A.: Ann. N.Y. Acad. Sci. *68*, 1254 (1958).
105) Haddow, A.: Ann. N.Y. Acad. Sci. *68*, 1258 (1958).
106) Ambrus, J. L., Ambrus, C. M., Back, N., Stutzman, L., Sokal, J. E., Ross, C. A.: The Pharmacologist *1*, 80 (1959).
107) Segaloff, A., Bardos, T. J., Papanastassiou, Z. B., Ambrus, J. L., Weeth, J. B.: Cancer Res. Proc. *3*, 62 (1959).
108) Dunning, W. F., Curtic, M. R., Armaghan, V., McKenzie, D.: Cancer Chemother. Rept. *12*, 1 (1961).
109) McCracken, S., Wolf, J.: Cancer Chemother, Rept. *6*, 52 (1960).
110) Razis, D. V., Ambrus, J. L., Ross, C. A., Stutzman, L., Sokal, J. E., Rejali, A. M., Bardos, T. J.: Cancer *14*, 853 (1961).
111) Rosenstiel, H., Bovil, E., Shabart, E., Scott, J. L., Yee, J. A., Lipschultz, B. M., Mass, R. F., Morris, J. F., Stevenson, J. A., Willett, F. M., Foye, L. L., Brandborg, O. S., Hayward, R. W., Phillips, R. L.: Cancer Chemother. Rept. *33*, 15 (1963).
112) Weeth, J. B., Segaloff, A.: Southern Med. J. *54*, 39 (1961).
113) Bardos, T. J., Chmielewicz, Z. F., Navada, C. K.: J. Pharm. Sci. *54*, 399 (1965).
114) Bardos, T. J., Ambrus, J. L.: 3rd Internat. Congr. Chemotherapy Vol II, p. 1036. Stuttgart: Georg Thieme Verlag 1964.
115) Bardos, T. J., Chmielewicz, Z. F., Hebborn, P.: Ann. N.Y. Acad. Sci. *163*, 1006 (1969).
116) Chmielewicz, Z. F., Fiel, R. J., Bardos, T. J., Ambrus, J. L.: Cancer Res. *27*, 1248 (1967).
117) Ross, C. A., Ambrus, J. L., Sokal, J. E., Velasco, H. A., Stutzman, L., Razis, D. V.: Cancer Chemother. Rept. *18*, 27 (1962).
118) Watne, A., Moore, G. E., Ambrus, J. L.: Cancer Chemother. Rept. *16*, 421 (1962).
119) Ross, C. A., Velasco, H. A., Sokal, J. E., Ambrus, J. L., Stutzman, L., Webster, J., Bardos, T. J.: Cancer Res. Proc. *3*, 355 (1962).
120) Regelson, W., Pierucci, O.: Proc. Am. Assoc. Cancer Res. *4*, 56 (1963).
121) Velasco, H. A., Ross, C. A., Webster, J. H., Sokal, J. E., Stutzman, L., Ambrus, J. L.: Cancer *17*, 841 (1964).
122) Webster, J. H., Sokal, J., Ross, C. A., Velasco, H. A., Ambrus, J. L., Stutzman, L.: Cancer Bull. *22*, 59 (1970).
123) Bardos, T. J., Ambrus, J. L., Ambrus, C. M.: J. Surg. Oncol. *3*, 431 (1971).
124) Szantay, C., Chmielewicz, Z. F., Bardos, T. J.: J. Med. Chem. *10*, 101 (1967).
125) Bardos, T. J., Datta-Gupta, N., Hebborn, P.: J. Med. Chem. *9*, 221 (1966).
126) Bardos, T. J., Baranski, K., Chakrabarti, P., Kalman, T. I., Mikulski, A. J.: Proc. Amer. Assoc. Cancer Res. *13*, 359 (1972).
127) Szabo, L., Kalman, T. I., Bardos, T. J.: J. Org. Chem. *35*, 1434 (1970).
128) Mikulski, A. J., Bardos, T. J., Chakrabarti, P., Kalman, T. I., Zsindely, A.: Biochim. Biophys. Acta *319*, 294 (1973).
129) Chandra, P., Bardos, T. J., Chakrabarti, P., Ebener, U., Ho, Y. K., Mikulski, A. J., Zsindely, A.: in preparation.
130) Chandra, P., Bardos, T. J.: Res. Commun. Chem. Pathol. Pharmacol. *4*, 615 (1972).
131) Srivastava, B. I. S., Bardos, T. J.: Life Sci. *13*, 47 (1973).
132) Bhargava, P. M., Shanmugan, G.: Prog. in Nucl. Acid Res. and Mol. Biol. *11*, 104 (1971).
133) Busch, H., Choi, Y. C., Crooke, S. T., Okada, S.: Oncology *26*, 152 (1972).

Molecular Approaches for Designing Antiviral and Antitumor Compounds

Professor Dr. Prakash Chandra

Klinikum der Johann Wolfgang Goethe-Universität, Gustav-Embden-Zentrum der Biologischen Chemie, Abteilung für Molekularbiologie, Frankfurt (Main)

Contents

P. Chandra

Introduction

Information transfer in biological systems usually involves transfer from DNA
to DNA (DNA replication), DNA to RNA (transcription) and RNA to protein
(translation). Most RNA viruses have an additional mode of passing information:
RNA to RNA (RNA replication). RNA tumor viruses have yet another additional
way of passing information: RNA to DNA (reverse transcription).

The discovery of reverse transcriptase in oncogenic RNA viruses[1, 2] and
human leukemic cells[3] opens a new horizon for the study of the role of viruses
in cancer. It may eventually enable us to design useful drugs for the selective
chemotherapy of cancer. The RNA-dependent DNA polymerase, or reverse
transcriptase, of virions is responsible for the synthesis of DNA chains on the
RNA template, which give rise to a hybrid molecule (RNA–DNA). These
chains are released from the RNA template as single-stranded DNA molecules
and serve as the template for the synthesis of double-stranded DNA (Fig. 1).

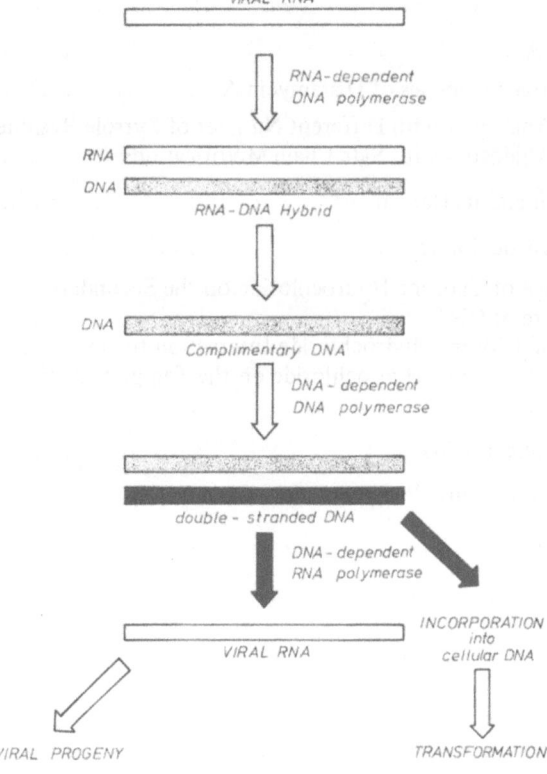

Fig. 1. Schematic representation of DNA synthesis in RNA tumor viruses

100

During the last decade, considerable progress has been made in the chemotherapy of cancer. This is a highly specialized field in which hundreds of investigators from various scientific disciplines cooperate. As a result of this tremendous scientific activity, numerous compounds have been synthesized and screened for antitumor activity; a few have proven to be useful against some types of human cancer. These compounds are mostly of an antimetabolic nature, such as purine and pyrimidine analogues and their cytotoxicity has precluded their use in human patients. The group of antitumor antibiotics is unfortunately no exception to this generalization, as exemplified by mitomycin and actinomycin D which, though effective against some human tumors, are highly cytotoxic. It is therefore desirable to look for new compounds in which a low cytotoxic effect accompanies high antitumor activity.

Cytotoxic activity, like cytostatic and antiviral activity, requires specific structural entities in a molecule[4−7]. All these activities may be derived from the same structural moiety, or various parts of the molecule may be responsible for them. Thus, structural modifications in an antitumor compound are extremely useful for

a) studying its mode of action, and

b) developing derivatives showing promise for the future chemotherapy of cancer. We report the role of chemical structure in the inhibition of DNA polymerases from RNA tumor viruses, viral multiplication and tumor growth by some antitumor compounds and their structural analogues.

1. Distamycin A

Distamycin, a mixture of antibiotic substances exhibiting predominantly antifungal activity, was obtained by submerged fermentation and butanol extraction of the mycelial mass of a *Streptomyces* sp. An amorphous product (frac-

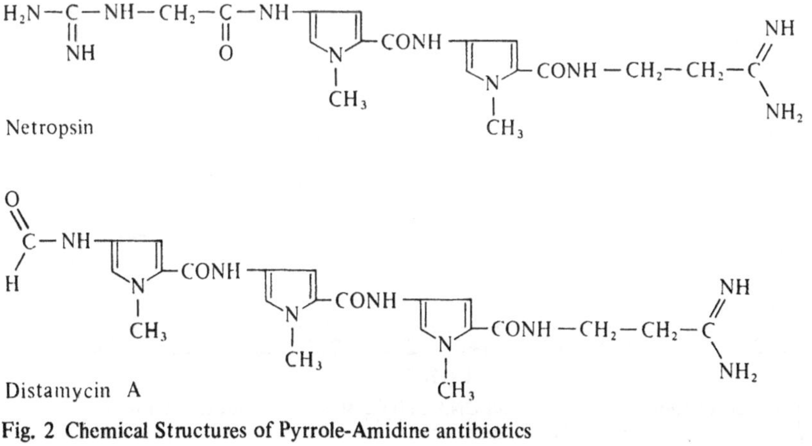

Fig. 2 Chemical Structures of Pyrrole-Amidine antibiotics

tion A), showing a positive Ehrlich reaction and a typical ultraviolet spectrum, was separated from the other active components by solvent fractionation and chromatography on aluminium oxide[11]. Partially purified preparations of this fraction, obtained during isolation investitions, were found to inhibit in various degrees some experimental tumors in mice[12] and to interfere with the process of cell division in vitro[13]. The ultraviolet and infrared spectra of these preparations showed some similarities with the antibiotic netropsin. Chemical investigations[14] indicated that the structure of *distamycin A* is characterized by three residues of 1-methyl-4-aminopyrrole-2-carboxylic acid and two side chains (Fig. 2).

Other compounds which have been isolated from the fermentation broths of *S. disthallicus* are netropsin (Fig. 2) and a simple pyrrole derivative devoid of biological activity. Netropsin (synonium congocidin) was reported in 1951[15],

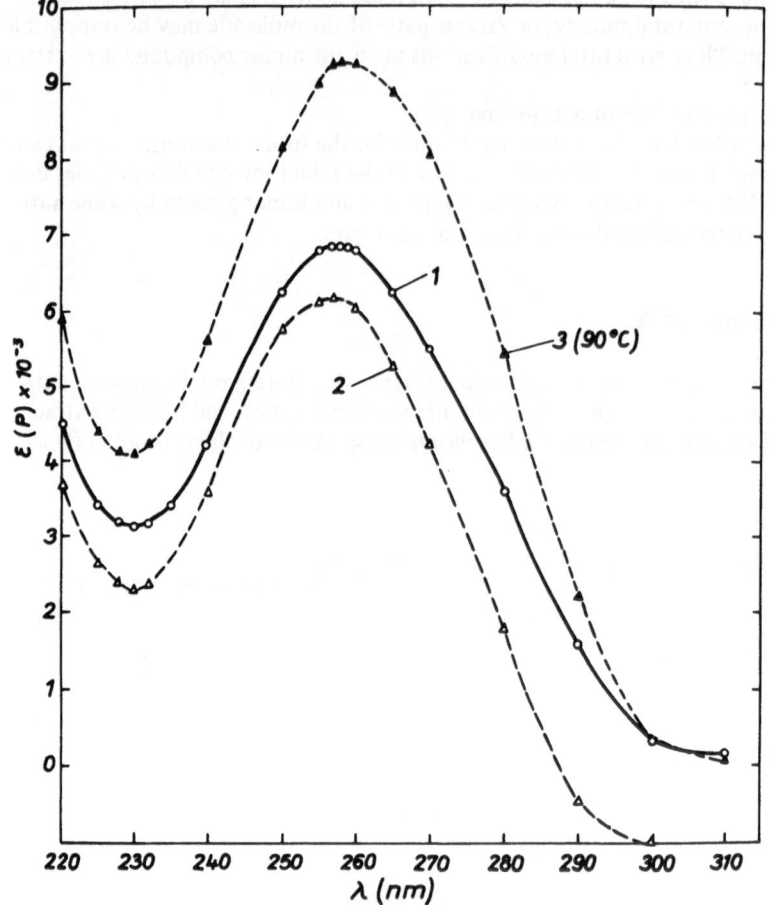

Fig. 3. UV-absorption spectra of *E. coli* DNA in 0.02 M KCl without (*1*) and with (*2*) distamycin A at 25 °C and 90 °C. Chandra *et al.*[22]

and its structure was definitely established by Julia and Joseph[16]. Netropsin appeared to be primarily active against both gram-positive and gram-negative bacteria, and against protozoa, but the compound is also endowed with antiviral properties [17, 18].

The influence of distamycin A on the UV-absorption of DNA is shown in Fig. 3. The absorbance of DNA decreases in the presence of distamycin A. This effect is dependent on the antibiotic/DNA-P ratio (r). After thermal denaturation, the absorbance reaches a constant value.

The ultraviolet absorption spectra of the distamycin-DNA systems beyond 300 nm at ionic strength of 0.01, pH 7.0 has also been studied[19]. When increasing amounts of DNA were added to the distamycin solution, the absorption maximum was shifted from 303 nm to longer wavelengths. The redshifts depended on DNA concentration. At a distamycin/DNA-P molar ratio close to 0.1, native DNA caused a red-shift of the absorption maximum of about 18 nm. These spectral changes were not inhibited by 10^{-1} M magnesium ion. A similar effect of DNA on the absorption spectrum of distamycin was observed by Zimmer et al.[20] and by Krey et al.[21]. These spectral changes have been interpreted in terms of a contribution of the system of the chromophore to the binding process with DNA.

The effect of the oligopeptide antibiotic on the helix-coil transition of DNA is demonstrated by thermal melting of the DNA-distamycin complex. As shown in Fig. 4, the melting profile of native DNA shifts towards higher temperatures with increasing antibiotic concentration. The hyperchromicity

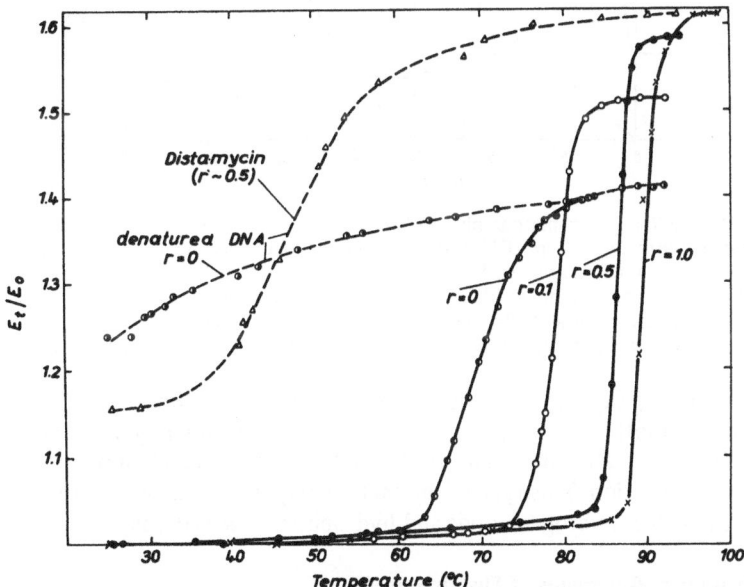

Fig. 4. Melting temperature of native and denatured calf thymus DNA in 0.02 M KCl in the presence of various distamycin A/DNA-P ratios (r). Chandra et al.[22]

also increases from 40 to about 60% when r is raised from $r = 0$ to $r = 1$. It is interesting to note that denatured DNA shows a cooperative transition and a hyperchromicity of 60% in the presence of distamycin A. From these results we conclude a strong binding of distamycin A to DNA.

The exact mode of binding of distamycin A to DNA has been further studied by Chandra et al.[23] using DNA-cellulose columns. The DNA-cellulose column chromatography was carried out as described by Inagaki and Kageyama[24]. Fig. 5 shows the chromatographic behaviour of distamycin A on a DNA-cellulose column. The unbound distamycin was completely removed by washing the column thoroughly with a standard buffer containing no NaCl. Using the gradient elution technique, the elution pattern of compound I showed two peaks centered at about 0.25 M–1 M in NaCl gradient and at about 3 M–4.5 M in urea gradient respectively. Similar behaviour was found, when the elution was carried out stepwise. The recovery was almost quantitative. The purity of the compound was checked by thin-layer chromatography. It is, therefore, unlikely that the elution pattern can be attributed to a mixture of two different molecular species.

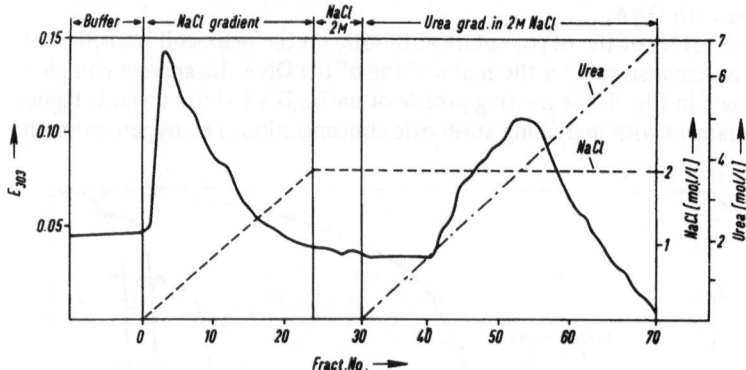

Fig. 5. DNA-cellulose chromatography of distamycin A. An excess amount of distamycin A was loaded into a 0.6x4 cm DNA-cellulose column. After washing the column with the buffer, elution was carried out with a linear gradient (0–2M) of NaCl in 100 ml of 0.01 M Tris-HCl (pH 7.0) containing 0.001 M EDTA, followed by 30 ml of 2M NaCl, and subsequently with a linear gradient (0–7 M) of urea in 200 ml of former eluent. This experiment was carried out at room temperature. Flow rate 36 ml/h. Recovery was 96%

This finding suggests the occurrence of at least two modes of interaction between antibiotic molecules and DNA; an electrostatic binding mode involving probably the DNA phosphate groups and a stronger binding mode requiring urea for dissociation. This type of interaction may involve some other kinds of forces, such as hydrogen bonds and hydrophobic associations.

1.1. Structural Analogues of Distamycin A

Synthetic analogues of distamycin A may be divided into two major classes.

To the first class belong those analogues which, unmodified in the side chain, contain a different number of residues of 4-amino-1-methylpyrrole-2-carboxylic acid. In the second class are considered those analogues which show modifications in side chains of the distamycin molecule. These modifications have been achieved either in the propionamidine side chain, or in the formyl-amino side chain. These analogues have been synthesized by Arcamone *et al.*[25, 26].

1.1.1. Analogues with Different Number of Pyrrole Residues

The initial compound of the series (n = 1) was obtained[26] from β-(1-methyl-4-nitropyrrole-2-carboxamido)-propionitrile. This compound is completely devoid of biological, as well as biochemical activity. Compounds containing at least two (n = 2) or more pyrrole groups are of great interest to study structure-activity relationship in various systems. The present study describes the biochemical basis of action of distamycin analogues containing two or more pyrrole groups (Fig. 6).

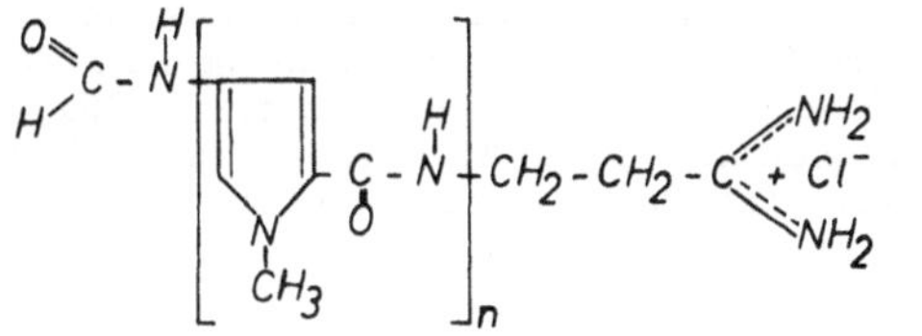

n = 2	Dist/2
n = 3	Dist/A
n = 4	Dist/4
n = 5	Dist/5

Fig. 6. Distamycin A analogues with different number of pyrrole residues

Fig. 7 compares the effects of native and denatured DNA, and of RNA on the ultraviolet spectra of the oligopeptide antibiotics containing 3, 4 and 5 pyrrole rings. Three general observations contained in this figure need emphasis.

First, yeast RNA and (in the case of distamycin-A) apurinic and apyrimidinic DNA at r = 0.025 did not change the characteristic absorption band of the free compounds. The absorption spectrum of distamycin-5 was enhanced in the presence of RNA.

Second, heat-denatured DNA at pH 7.0 bound all the compound tested and, at corresponding values of "r", the spectra of the bound antibiotics were similar to those found with native DNA.

Finally, in each case, the red shift of the characteristic absorption band was accompanied by an increase in the intentsity of the absorption maximum. This hyperchromic effect depended upon DNA concentration and became more pronounced as the number of pyrrole residues in the distamycin molecule increased. Under identical conditions of "r" and ionic strength, the spectral changes induced by denatured DNA were somewhat weaker.

The differences in hyperchromic effect can hardly be attributed to differences in the stability of the compounds tested, as the spectra of the compound both in the absence and in the presence of DNA were read immediately after preparation.

These observations provide strong support for the view that the chromophore of the antibiotic contributes to the binding mechanism.

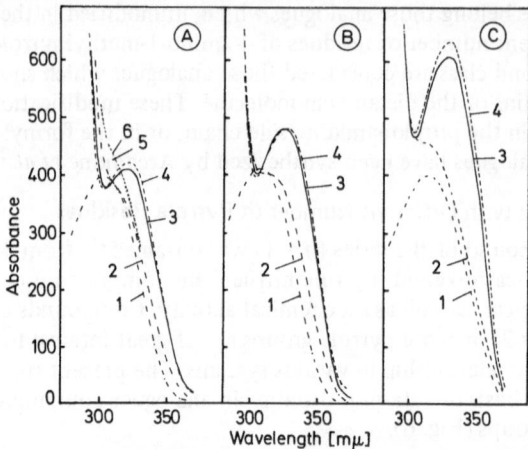

Fig. 7. Effect of native DNA, denatured DNA and yeast RNA on the ultraviolet absorption spectra of distamycin (A), distamycin/4 (B) and distamycin/5 (C), in 0.01 M Tris-HCl buffer (pH 7.0). Curve *1* is the spectrum of free antibiotics. Other curves are spectra of the antibiotics in the presence of yeast RNA (curve *2*), denatured DNA (curve *3*), native DNA (curve *4*), apurinic acid (curve *5*) and apyrimidinic DNA (curve *6*) at a molar antibiotic/nuclei acid-P ratio of 0.025. Chandra *et al.*[27]

Concerning the relationship between the above contribution and the number of pyrrole residues, an attempt was made to investigate whether the extent of binding and/or the affinity for DNA could be affected by the number of pyrrole residues. The results obtained by the displacement of methyl green from its DNA complex by distamycin, reported by Krey and Hahn[21], suggested that methyl green and distamycin attach to the same binding site of DNA. We observed similar displacement reaction with distamycin/4 and distamycin/5. In addition, a comparison of the rates of displacement suggested an increase in relative affinity for DNA of the compound distamycin/5 with respect to distamycin. Furthermore, an increase in the ability to displace methyl green by increasing the number of pyrrole residues was abserved, indicating an increase in the extent of binding of displacing compounds to DNA.

These data are in good agreement with those of Zimmer *et al.*[28] and Chandra *et al.*[29], who found that the stabilizing action of distamycin on DNA increased with increasing number of pyrrole residues.

Table 1. Cytotoxicity and antiviral activity of the distamycin derivatives

Compound	Pyrrole rings	% Inhibition of vaccinia virus multiplication[1]	Cytotoxicity[1]
Distamycin/2	2	13	100
Distamycin/A	3	100[2]	100[2]
Distamycin/4	4	400	50
Distamycin/5	5	1000	25

[1] Activity calculated with respect to distamycin A considered = 100.
[2] Absolute values (ID_{50}mcg/ml): Cytotoxicity = 80; WR = 2.

The cytotoxicity of distamycin derivatives was estimated on the basis of the morphological modifications induced in HeLa cell cultures, after incubation for 40 h in Hanks' saline solution + 0.5% lactalbumin hydrolysate + 5% calf serum (HLS). Assay on vaccinia virus: Cultures of HeLa cells (grown in HLS medium) or mouseembryo cells (grown in HLS medium plus 0.1% yeastolate) infected with vaccinia virus (Strain WR/ATCC) were used. Preliminary assays were made according to Herrmann et al.[30]. Subsequent studies were carried out by assessing the inhibition of plaque formation (ECP) as well as the inhibition of infectious virus production in test tube cultures treated with the compounds for 40 h after the absorption of the virus.

The *cytotoxicity and antiviral activity* of distamycin derivatives containing 2, 3, 4 and 5 pyrrole rings is shown in Table 1. The cytotoxicity of the derivatives with 2 and 3 pyrrole residues is the same. However, compounds containing 4 and 5 pyrrole rings (distamycin/4 and distamycin/5) are less toxic. The cytotoxicities of distamycin/4 and distamycin/5 are only 50% and 25% of the natural antibiotic (distamycin/A) respectively. It seems therefore, that the cytotoxicity decreases as the number of pyrrole rings increases. This is at least true for distamycin/A, distamycin/4 and distamycin/5. Our studies on distamycin/6 (distamycin with 6 pyrrole rings) have, however, shown that no such relationship strictly exists. Distamycin/6 was found to be as toxic as distamycin/4.

The antiviral activity of distamycin derivatives is dependent on the pyrrole ring. Taking the antiviral activity of the natural antibiotic (distamycin/A) as 100, one observes a 4-fold increase for distamycin/4 and a 10-fold increase for distamycin/5. On the other hand we found about 85% inhibition of the antiviral activity of distamycin/A by removing 1 pyrrole ring (distamycin/2).

Table 2. Distamycin A inhibition of DNA-dependent RNA polymerase reaction. Distamycin A was pipetted into reaction mixture containing calf thymus DNA, buffer and the triphosphates. The reaction was started with DNA-dependent RNA-polymerase

System	AMP-^{3}H Incorporation cpm/reaction mixture	% Incorporation
Complete	373	100
Without DNA	11	3
Complete + distamycin A		
4×10^{-5}M	111	29.6
8×10^{-5}M	60	16.1

RNA-polymerase reaction: RNA-polymerase was isolated from *E. coli* K$_{12}$ cells according to the procedure by Burgess[31] and kept in buffer containing 50% glycerol at -20 °C. The reaction mixture contained, in 0.25 ml, 0.04 M Tris, pH 7.9, 0.01 M MgCl$_2$, 0.1 mM EDTA, 0.1 mM dithiothreitol, 0.15 M KCl, 0.15 mM UTP, CTP and GTP, 0.15 mM ^{3}H-ATP and 0.15 mg per ml of calf thymus DNA. The reaction was started with 5–10 mcg enzyme protein and incubations were carried out for 20 min at 37 °C.

Using the *melting behaviour* of DNA-antibiotic complexes as a criterion of binding a drastic increase in the melting temperature of DNA was observed in the presence of distamycin A[22]. This interaction leads to a concentration-dependent inhibition of DNA-dependent RNA polymerase reaction. Table 2 shows the template activity of calf thymus DNA in the presence of the natural antibiotic (distamycin/A). In these experiments distamycin/A was pipetted into reaction mixture containing DNA, buffer and the triphosphates. The reaction was started with DNA-dependent RNA polymerase. One gets about 70% inhibition at 4×10^{-5} and 84% at 8×10^{-5} M. This is in good agreement with our previous results[22]. The effect of an equimolar concentration (4×10^{-5}M) of various distamycin derivatives on the template activity of calf thymus DNA is shown in Table 3. The derivatives were added into the reaction mixture as described above. The inhibition of DNA-dependent RNA synthesis increases as the number of pyrrole residues in the antibiotic molecule increases. The compound with 2 pyrrole rings inhibits ^{3}H-AMP incorporation to 50%, whereas distamycin/5 at the same molar concentration to 82%.

Table 3. Inhibition of DNA-dependent RNA polymerase reaction by distamycin derivatives

Compound added Concentration = 4×10^{-5}M	Pyrrole rings	AMP–^{3}H Incorporation cpm/reaction mixture	% Incorporation
None	–	373	100
Distamycin/2	2	187	50
Distamycin/A	3	111	29.6
Distamycin/4	4	79	21.2
Distamycin/5	5	67	18.0

It is known that RNA oncogenic viruses require DNA synthesis for their replication. As distamycin/A blocks some early steps in the growth cycle of DNA viruses, probably connected with DNA replication[32], it was of interest to investigate the effect of distamycin, distamycin-4 and distamycin-5 on MSV (Moloney).

The results, reported in Fig. 8, show that treatment with distamycin/A, distamycin/4 and distamycin/5 reduces the number of MSV-foci produced in vitro. The inhibitory activity was dependent on the dose used, and increased according to the number of pyrrole residues in the molecule. The cytotoxic activity of these compounds increases at higher doses; however, it was always less than their antiviral activity.

Since the dose-dependent curves for cytotoxicity and antiviral activity are almost parallel, it was possible to calculate the ratio cell-ID_{50}/virus-ID_{50}. As shown in Table 4, the Therapeutic index (= TI) increased from 2.5 for distamycin/A, to 7.0 for distamycin/4 and to 7.7 for distamycin/5.

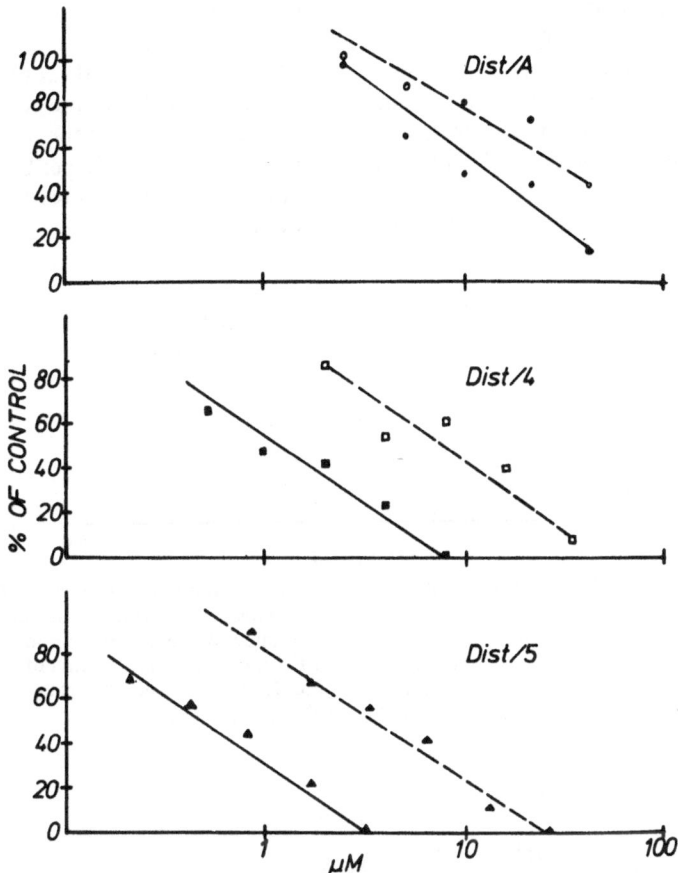

Fig. 8. Activity of distamycin derivatives on mouse embryo cells infected or not with MSV (M). (——) indicates inhibition of MSV-M foci formation; (----) inhibition of normal cell proliferation. On the abscissa: dose (μM); on the ordinate: % of the corresponding value. Chandra et al.[6]

Table 4. Antiviral and cytotoxic activities of some distamycin derivatives. The activity of distamycin derivatives was tested on mouse embryo cells infected or not with MSV (M). ID_{50} = Inhibiting dose 50%; TI = therapeutic index, ID_{50}-cell/ID_{50}-virus

Compound	ID_{50} (μM)		TI
	Cell	MSV	
Distamycin/A	32.0	12.6	2.5
Distamycin/4	7.8	1.1	7.0
Distamycin/5	3.4	0.44	7.7

To find whether the reduced *foci formation* was due to inhibition of MSV (M) replication, the effect of these compounds on virus growth was studied. Secondary mouse embryo cell cultures were infected with MSV (M) at the rate of 0.03 competent MSV infectious units per cell and treated with the compounds at different doses after the infection by the focus assay in the presence of an optimal amount of MLV (M) (1.01 · 10^5 Leukemia Virus Helper Units) according to Hirschman et al.[33]. Similarly treated uninfected cultures were trypsinized at the same time, and cells were counted.

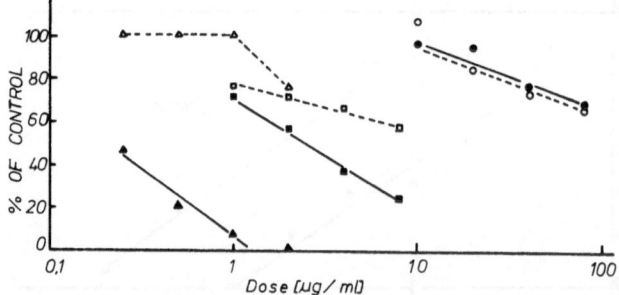

Fig. 9. Inhibition of MSV (M) replication and of cell proliferation by distamycin (○●), distamycin/4 (□■) and distamycin/5 (△▲). Empty symbols and continuous lines: inhibition of MSV (M) production (titeration 3 days after infection). MSV (M) titer in untreated cultures ranged from 3.0×10^5 to 6.1×10^5 FFU/ml. Filled symbols and dotted lines: inhibition of secondary mouse embryo cell proliferation. Cell number/plate in untreated cultures ranged from 1.1×10^5 to 2.6×10^5. Chandra et al.[27]

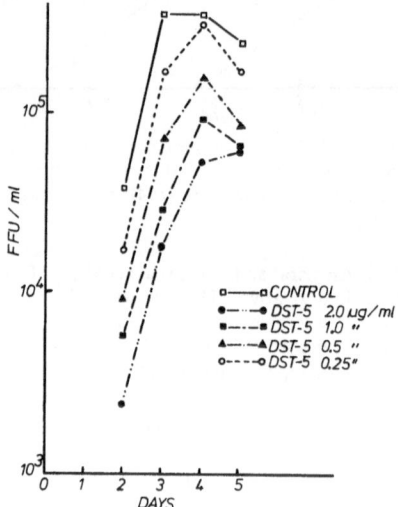

Fig. 10. Growth curve of MSV(M) in secondary mouse embryo cell cultures, untreated or treated with distamycin/5. The titers represent the total virus [yield at each day, determined with added excess helper virus]. Each day the cells were harvested in the supernatant and samples of this suspension were used for titration of MSV(M). Chandra et al.[27]

Fig. 9 shows the results regarding titration of virus yields and of cell proliferation on the 3rd day after infection, which is the day of maximum virus yield in our experimental conditions. Distamycin-A, even at cytotoxic doses, poorly inhibited the virus replication; distamycin-4 was very active on virus replication and distamycin-5 at non-cytotoxic doses was able to reduce viral yields by 92.3%.

Fig. 10 shows the viral yields obtained on days 2, 3, 4 and 5 after the infection, in control cultures and in cultures treated with distamycin-5. The inhibition of virus production was almost constant on any day tested, suggesting that an early step in virus multiplication is blocked. Similar results were obtained with distamycin-4, but virus inhibition was less than with distamycin-5.

In mice distamycin-4, administered intraperitoneally 6 times/day for 6 days, starting 2 days after the infection, was able to inhibit and to reduce the growth of tumors induced by giving MSV (M) intramuscularly. Good results were also

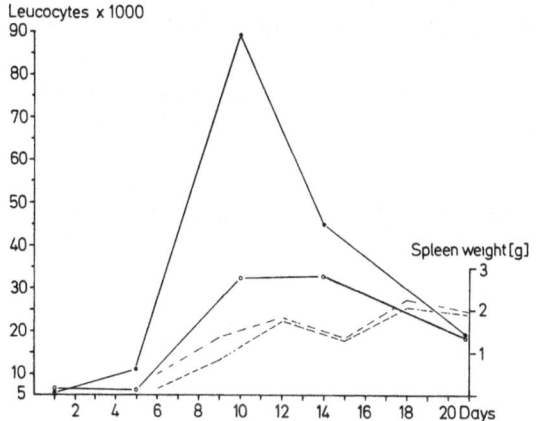

Fig. 11. Effect of distamycin A on leukemogenesis in mice by FLV. Viral suspensions, untreated (ID_{90}) or treated with distamycin/A (20 μg/ml) were injected intraperitoneally to mice. Continuous lines (filled circles) represent the number of leucocytes in mice treated with control suspension; open circles, leucocyte number in mice injected with distamycin/A treated suspension. The dashed line (with dots) represents the spleen weight of animals treated with the control suspension; without dots, the spleen weights of animals injected with distamycin/A treated suspension

obtained by administering distamycin, distamycin-4 and distamycin-5 intramuscularly in the MSV (M)-infected leg. Less activity was observed in vivo than in vitro. This may be due to uneven distribution in the animal body, metabolism and/or rapid elimination (G. Di Fronzo and G. Lenaz, unpublished results). Further studies will investigate the possibility of obtaining a more significant activity against oncogenic viruses in the animal by varying schedule and route of treatment.

The in-vivo activity of distamycin-A on *FLV leukemogenesis* in mice is shown in Fig. 11

The studies were carried out on male and female albino mice weighing 20–30 g. FLV suspensions were prepared by filtering the homogenates from infected spleens through Seitz K-filters (Seitz Company, Bad Kreuznach, Germany). This suspension was diluted 1 : 20 with Hanks solution and 0.1 ml (ID_{90}) was injected intraperitoneally. In the experimental group the suspensions were preincubated with distamycin-A (20 μg/ml), in the control group they were not; each group contained 50 animals. At various intervals 10 animals from each group were sacrificed and their leucocyte number, and spleen weight determined.

The results (Fig. 11) show a very significant reduction in the number of leucocytes in the distamycin-A treated group and a slight reduction in spleen weight, as compared to controls.

The *mechanism* by which distamycin-A and its structural analogues exert their activity against MSV (M) and FLV could be explained by inhibition of the reverse transcriptase activity. The existence of virion-associated DNA polymerases[1, 2] in oncogenic RNA viruses suggests that the flow of information from RNA to DNA may be one of the factors in oncogenesis. The search for specific inhibitors of this reaction has obvious implications for the chemotherapy of viral cancers. The RNA-dependent DNA polymerase can use as template synthetic polymers containing either deoxyribonucleotide or ribonucleotide strands. The activity of distamycin derivatives on the DNA polymerase activities of FLV and MSV (M) was tested using various templates.

Table 5. Inhibition of reverse-transcriptase activity of FL-virions by distamycin derivatives in the absence of exogenous template. Virions containing Triton were preincubated at room temp. for 25 min with 50 μg/ml of pancreatic RNase Chandra *et al.*[6]

System	Antibiotic 20 μg/reaction mixture	^{3}H–TMP incorporation into DNA (c.p.m./reaction mixture	% of control
Without virions	–	14	4.1
Without Triton	–	40	11.7
Virions + RNase	–	61	17.6
Complete	–	343	100
Complete	Distamycin-2	336	98
Complete	Distamycin-A	242	70.5
Complete	Distamycin-4	203	59
Complete	Distamycin-5	206	59.8

As shown in Table 5, the *reverse transcriptase activity* (without exogenous template) of FL virions is inhibited by distamycin derivatives containing 3, 4 and 5 pyrrole rings. The compound containing 2 pyrrole rings (distamycin-2) is ineffective, which confirms our observations on vaccinia virus. The compound with 4 pyrrole rings inhibits the reverse-transcriptase reaction more than the one with 3 pyrrole rings (distamycin-A). Further progression could not be demonstrated with the compound containing 5 pyrrole rings (distamycin-5) which is very unstable; longer incubation (90 min) may cause some degradation of this compound.

Table 6. Inhibition of DNA-polymerase activity of FL virions by distamycin A and its analogues in the presence of various templates. Figures in brackets are percentages

Antibiotic (20 μg/reaction mixture	^{3}H–TMP incorporation[1]) into DNA poly (dA–dT)	^{3}H–dGMP incorporation[1]) into DNA poly (dI-dC)	^{3}H–TMP incorporation[1]) into DNA poly (rA) · (dT)$_8$
None	1020 (100)	506 (100)	520 (100)
Distamycin-2	357 (35.8)	441 (87.5)	450 (85.5)
Distamycin-A	137 (14)	442 (87.5)	330 (63.5)
Distamycin-4	117 (11.9)	460 (91)	290 (55.5)

[1]) cpm/reaction mixture.

It has been recently reported[34] that DNA polymerases of several oncogenic viruses are inhibited by ethidium bromide to different degrees according to the nature of the template used and the source of the enzyme. The effect of distamycin derivatives on the DNA polymerase activity of FL virions was therefore studied in the presence of poly (dA–dT), poly (dI · dC) and poly (rA) · (dT)$_8$ (Table 6). The DNA polymerase activity was found to be most sensitive to distamycin inhibition with poly (dA–dT) as primer-template and somewhat less so with poly (rA) · (dT)$_8$. In both cases the inhibitory response of the antibiotic increases according to the number of pyrrole rings in the molecule. With poly (dI · dC) as template no significant inhibition of DNA polymerase activity by distamycin derivatives was observed.

Table 7. Inhibition of DNA polymerase activity of MSV (M) by distamycin A and its derivatives in the presence of various templates. The incorporation in the absence of templates was 157 cpm/reaction mixture for ^{3}H–TMP, and 61 cpm/reaction mixture for ^{3}H–dGMP. – Antibiotic concentration = 20 μg/reaction mixture. The figures in brackets indicate percent of the control

Antibiotic	^{3}H–TMP incorporation into DNA poly (dA-dT) (cpm/reaction mixt.)	^{3}H–dGMP incorporation into DNA poly (dI-dC) (cpm/reaction mixt.)	^{3}H–TMP incorporation into DNA poly rA · (dT)$_8$ (cpm/reaction mixt.)
None	624 (100)	570 (100)	412 (100)
Distamycin/2	204 (32.6)	459 (86)	328 (80)
Distamycin/A	144 (23)	441 (77)	280 (68.2)
Distamycin/5	60 (9.6)	447 (78)	214 (52)

Inhibition of DNA polymerase activity of MSV-M by distamycin A and its derivatives in the presence of various templates is shown in Table 7. The DNA-polymerase activities catalyzed by poly (dA-dT) and poly rA-(dT)$_8$ are very sensitive to distamycin action. The range of inhibitions by individual derivatives is similar to that observed in FL-virions. However, the poly (dI-dC) dependent incorporation of dGMP by DNA polymerase was found to be more sensitive towards distamycins in this case.

The experiments reported here demonstrate that the distamycin inhibition of DNA polymerase activities of FLV and MSV-M are template specific. Templates containing thymine and adenine are highly sensitive to the action of distamycins. This inhibition is dependent on the number of pyrrole rings in the molecule. The inhibition of DNA polymerases of RNA oncogenic viruses and the foci formation by distamycin derivatives conclude that both activities are dependent on the same structural component(s) of the molecule.

1.1.2. Analogues with Side Chain Modifications

The cytotoxicity and antiviral activity of distamycin derivatives, obtained by substitutions of the formyl group (II) or the propionamidine chain (III and IV),

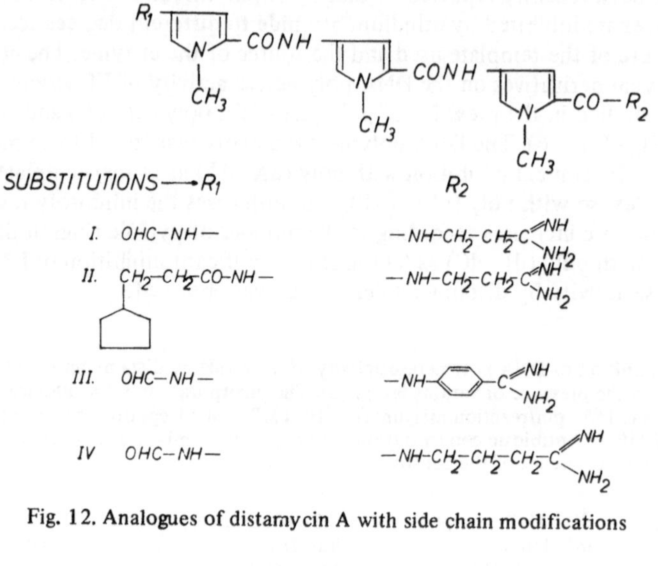

Fig. 12. Analogues of distamycin A with side chain modifications

is shown in Table 8. Substitution of the formyl group with a cyclopentyl propionyl chain does not influence its cytotoxicity but the compound loses its antiviral activity completely. The substitution of the propionamidine group with a benzamidine moiety doubles the cytotoxicity of the compound. The antiviral activity of this compound is, however, only 44% of that of distamycin A. The analogue containing butyramidine group in place of the propionamidine moiety (IV) also exhibit a higher cytotoxicity than the compound I. This has only 31% of the antiviral activity of compound I.

These results allow the conclusion that the *presence of the formyl group in distamycin is necessary* for its antiviral activity. Studies with other derivatives, where the formyl group was substituted by a nitro, amino or acetyl group have

Table 8. Cytotoxicity and antiviral activity of the distamycin derivatives

Compound tested	Cytotoxicity[1]	Inhibition of vaccinia virus multiplication. inhibition (%)[1]
I	100[2]	100[2]
II	100	0
III	200	44
IV	150	31

[1]) Activity calculated with respect to that of compound I (distamycin A) considered = 100.
[2]) Absolute values (ID_{50} μg/ml): Cytotoxicity = 80; WR = 2.

shown that all these derivatives are completely inactive against viruses. However, substitutions at the formyl group do not influence the cytotoxicity of the compound. Compounds having substitutions at the propionamidine are active against viruses but exhibit a much higher toxicity. An interesting compound of this group is the acetamidine derivative which showed a higher antiviral activity (150%) than distamycin A.

Table 9. Inhibition of DNA-dependent RNA polymerase reaction by distamycin derivatives

Compound added (concentration = 8×10^{-5}M)	AMP–^{3}H incorporation (cpm/reaction mixture)	Incorporation (%)
None	3801	100
I	650	17
II	1792	47
III	1990	53
IV	980	26

Chandra *et al.*[4] have shown that the antiviral activity of distamycin and their action on the template activity of DNA are dependent on the *number of pyrrole rings* in the molecule. The distamycin derivative with 5 pyrrole rings (distamycin/5) has 10 times higher antiviral activity than distamycin A (with 3 pyrrole rings), and is a better inhibitor of the RNA-polymerase reaction. It was therefore interesting to study the correlation between the antiviral activity and the inhibition of RNA-polymerase reaction by compounds I, II, III and IV. The effect of an equimolar concentration (8×10^{-5} M) of the distamycin derivatives on the template activity of DNA is shown in Table 9. The derivatives were added into the reaction mixture as described above. The highest inhibition was obtained with the natural antibiotic (distamycin A) followed by compounds IV, II and III, respectively. The cyclopentylpropionyl derivative, having no antiviral activity is still able to inhibit the RNA-polymerase reaction to more than 50%. This indicates that factors, other than its binding to DNA, are responsible for its inactivity against viruses. One of the many possibilities for this result may be the permeability of this compound towards the host cell. The inhibition of RNA-polymerase reaction by compounds III and IV is in good correlation to their antiviral activity, compared to distamycin A.

To investigate the role of *formyl group* some amino acid derivatives of des-formyldistamycin A were synthesized, in which the formyl group was linked to the amino group of glycine or alanine. Thus, the side chains of these derivatives are constituted by N-formyl-glycyl, or N-formyl-alanyl groups. Studies on various systems reported here indicate that the compound with formyl group linked to 4-amino-1-methylpyrrole-2-carboxylic acid residue (distamycin/A) is more active compared to its amino acid derivatives.

The effect of distamycin/A and its amino acid derivatives (distamycin/Gly and distamycin/Ala) on the template activity of calf thymus DNA in RNA-

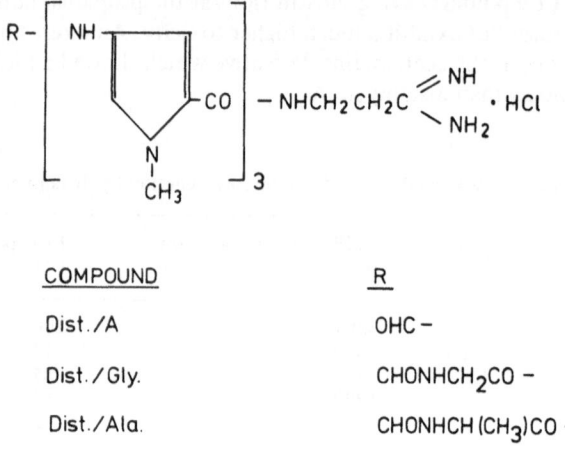

COMPOUND	R
Dist./A	OHC –
Dist./Gly.	$CHONHCH_2CO$ –
Dist./Ala.	$CHONHCH(CH_3)CO$ –

Fig. 13. Amino acid derivatives of des-formyl-distamycin A

polymerase (*E. coli*) reaction is shown in Table 10. In these experiments equimolar concentrations of the antibiotics (8×10^{-5} M) were added to the incubation mixture, and the reaction was started immediately with the enzyme. Under these conditions distamycin/A inhibits appx. 70% of the incorporation of ^{3}H–AMP into RNA. Distamycin/Gly and distamycin/Ala at the same molar concentrations are much less effective in inhibiting the template activity of DNA.

The above results indicate that the binding affinity of amino acid derivatives for DNA is much less, compared to that of the parent compound. Thus, the position of the formyl group is important in its binding mechanism to DNA. This has been further shown by studies on the DNA-polymerases of oncorna viruses and bacterial cells.

Table 10. Inhibition of DNA-dependent RNA polymerase reaction (*E. coli* B) by distamycin A and its structural analogues

System	Antibiotic (concentration = 8×10^{-5}M)	^{3}H–AMP incorporation into RNA (cpm/reaction mixt.)	% of Control
Without DNA	–	17	1.5
Without Enzyme	–	5	0.44
Complete	None	1144	100
Complete	Distamycin /A	346	30.3
Complete	Distamycin/Gly	624	54.4
Complete	Distamycin/Ala	780	68.2

Table 11. Inhibition of reverse-transcriptase activity of FL-virions by distamycin derivatives in the absence of exogenous template

System	Antibiotic (concentration = 2×10^{-4}M)	^{3}H–TMP incorporation into DNA (cpm/reaction mixt.)	% of Control
Without virions	–	11	3
Without Nonidet P-40	–	45	12.2
Complete	–	364	100
Complete	Distamycin/A	93	25.6
Complete	Distamycin/Gly	146	40
Complete	Distamycin/Ala	258	71

Table 11 shows the activities of distamycin/A and its amino acid derivatives on the DNA polymerase activity of *Friend leukemia virions* (FLV). These studies were carried out without the exogenous template. It has been emphasized that the study of endogenic reaction has greater implications in viral cancerogenesis. The results show a strong inhibition of the endogenic reaction by distamycin/A at 2×10^{-4} M; distamycin/Gly was also very effective, distamycin/Ala showed a poor activity.

The endogenic reaction is catalyzed by the viral RNA (70s), template for the reverse transcriptase. It is not clear whether the higher activity of distamycin/Gly in this system, compared to the template activity of DNA in bacterial system (Table 10), is due to its higher affinity for viral RNA. This was investigated by using various exogenous templates in the DNA-polymerase reaction of FL-virions.

Synthetic polymers containing either deoxyribonucleotide or ribonucleotide strands can be used as templates by the DNA polymerases of RNA tumor viruses. Table 12 shows the activity of distamycin/A and its amino acid derivatives on the DNA polymerase reaction of FL-virions, catalyzed by poly rA ·

Table 12. Inhibition of DNA-polymerase activity of FL-virions by distamycin A and its analogues in the presence of various templates

Antibiotic (concentration = 2×10^{-4}M)	^{3}H$-$TMP incorporation into DNA (cpm/reaction mixture) in the presence of	
	Poly rA $\cdot$ (dT)$_{12}$	Poly (dA-dT)
None	2517 (100)	1282 (100)
Distamycin/A	879 (34.9)	381 (29.6)
Distamycin/Gly	1440 (57.3)	502 (39.3)
Distamycin/Ala	1742 (69.3)	745 (58.3)

(dT)$_{12}$ and poly (dA-dT). The poly rA $\cdot$ (dT)$_{12}$-catalyzed reaction is inhibited by distamycin/A and distamycin/Ala to same extent as the endogenous reaction. However, distamycin/Gly is in this reaction, compared to the endogenous reaction, less effective. The reaction catalyzed by the DNA template, *i.e.* poly (dA-dT) is slightly more sensitive than the RNA-dependent reaction towards distamycin/A. Surprisingly, distamycin/Gly showed a strong inhibition of this reaction, compared to the reaction in bacterial cells (Table 10).

Table 13. Inhibition of DNA-dependent DNA polymerase reaction (*E. coli* B) by distamycin A and its structural analogues

System	Antibiotic (concentration = 1×10^{-4}M)	^{3}H$-$dAMP incorporation into DNA (cpm/reaction mixture)	% of Control
Without DNA	–	98	0.67
Without Enzyme	–	41	0.28
Complete	None	14,609	100
Complete	Distamycin/A	2,392	16
Complete	Distamycin/Gly	5,368	37
Complete	Distamycin/Ala	9,517	65

The activity of these derivatives on the DNA-polymerase reaction of bacterial cells, catalyzed by denatured DNA is shown in Table 13. These studies are important to compare the sensitivities of viral and bacterial DNA polymerases towards these antibiotics. As follows from results the DNA polymerase reaction of bacterial cells is highly sensitive to distamycin/A and distamycin/Gly. The molar concentration of the antibiotics used in this reaction (1×10^{-4}M) is slightly higher, compared to the DNA-dependent RNA-polymerase reaction (8×10^{-5}M); however, the inhibitory effect in the former reaction is much more pronounced.

The studies reported above demonstrate that the replacement of formyl group, linked to amino group of the side chain, by N-formyl amino acids diminishes their biochemical activity. It is interesting to note that the activity of distamycin/Gly is dependent on the enzyme system; the compound has a good activity in the endogenous viral system and DNA-polymerase reaction of bacteria. The activity of these derivatives on the viral cancerogenesis in vivo is currently being investigated.

2. Daunomycin and its Derivatives

Daunomycin is an antibiotic of the anthracycline group isolated from cultures of *Streptomyces peucetius*[35, 36] and consists of a pigmented aglycone (dauno-mycinone) bound by a glycosidic linkage to an amino sugar (daunosamine, Fig. 14)[37, 38]. The biological activity of daunomycin is believed to be related to its ability to interact with the primer DNA[38, 40], thus inhibiting not only DNA-dependent RNA synthesis, but also DNA duplication[41].

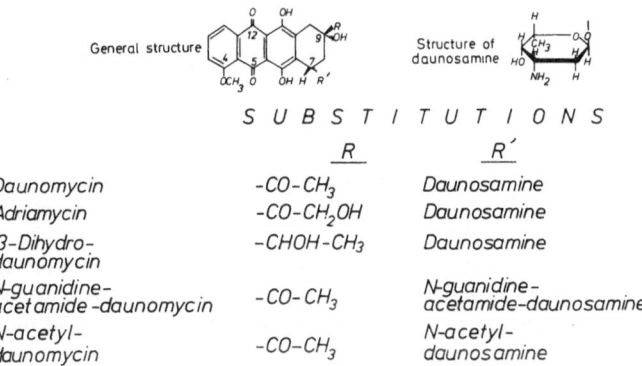

Fig. 14. Chemical structures of daunomycin derivatives

Daunomycin causes a large increase in the thermal transition temperature (T_m) of calf thymus DNA. This effect depends on the ratio antibiotic/DNA $\cdot$ P (r). The effect of daunomycin derivatives on the thermal transition temperature of calf thymus DNA at $r = 0.1$ is shown in Table 14. Adriamycin was found to be most effective in stabilizing the secondary structure of DNA ($\Delta T_m = 15.3\ °C$), whereas very little increase in T_m was observed for N-guanidine-acetamide dauno-mycin and N-acetyl daunomycin, the derivatives with substitutions in the amino sugar moiety. In attempting to obtain further information on the affinity of the compounds tested for DNA, we studied their effect on the viscosity of DNA. According to Lerman's hypothesis[42] on the interaction of amino-acridines with DNA, an increase in the intrinsic viscosity of the complex is one of the criteria establishing intercalation of ring systems between base pairs of double-stranded DNA. Table 14 shows the intrinsic viscosity of antibiotic-DNA com-plexes ($r = 0.1$) relative to intrinsic viscosity of DNA alone ($r = 0$). Under these

119

Table 14. Effect of daunomycin and its derivatives on the thermal transition temperature (T_m) and viscosity of DNA

Antibiotic	T_m[1])	$\dfrac{(\eta)\, r = 0.1\,[2])}{(\eta)\, r = 0}$
None	70.5	1.00
Adriamycin	85.8	1.75
Daunomycin	83.9	1.92
Dihydro daunomycin	80.3	1.65
N-guanidine acetamide daunomycin	77.1	1.30
N-acetyl daunomycin	71.5	1.24

[1]) All experiments were carried out in 0.01 M tris-HCl buffer (pH 7.0) at a ratio of antibiotic to DNA-P (r) of 0.1. DNA concentration in all the experiments was 1×10^{-4}M.
[2]) Ratio of intrinsic viscosity of antibiotic-DNA complex ($r = 0.1$) to that of DNA alone. Conditions of viscosity measurements: 20 °C, 0.1 M tris-HCl buffer (pH 7.0). r is the ratio of bound antibiotic to total DNA-P.

conditions the daunomycin-DNA complex has the highest intrinsic viscosity, followed by adriamycin and dihydro daunomycin. Again we find only a moderate increase in intrinsic viscosity in the presence of compounds substituted at the amino sugar residue.

The inhibitory effect of the various daunomycin derivatives on *viral oncogenesis by FLV and RSV* is shown in Table 15. FLV suspensions, prepared as described above, were incubated with and without the antibiotic (50 μg/ml) for 1 h at 37 °C and 0.1 ml of this suspension (ID_{90}) was injected intraperitoneally into mice. Each experimental group contained six animals; five of the six control animals died after 13 days of infection at which time all the animals

Table 15. Effect of daunomycin and its derivatives on viral oncogenesis in mice and chickens

Antibiotic	Oncogenesis in mice by FLV animals survived[1]) animals infected	Oncogenesis in chicken by RSV[2]) Mean survival (days)
None	1/6	12.3
Daunomycin	6/6	30[3])
Adriamycin	5/6	28.3
Dihydro daunomycin	3/6	14.2
N-guanidine daunomycin	1/6	13.6
N-acetyl daunomycin	0/6	12.0

[1]) 13 days after infection. For details see text.
[2]) Each experimental group contained six chickens. The viral suspension (1 : 10) was incubated with 50 μg/ml of the antibiotic at 37 °C for 1 h. Control suspension was incubated without the antibiotic. 0.1 ml of this suspension was injected intraperitoneally.
[3]) One animal was still alive at the time of writing this paper.

injected with daunomycin-treated viral suspension were still alive. Adriamycin was also effective, whereas dihydro daunomycin had only moderate activity. The derivatives substituted in the amino sugar moiety were ineffective. Oncogenesis by RSV in chickens was similarly inhibited by daunomycin and its analogues, mean survival time being prolonged from 12.3 days to 28.3 days by adriamycin. Daunomycin was even more effective but, the other derivatives had no significant effect.

Daunomycin and related derivatives were found to affect cell proliferation and MSV (M) foci production dif'erently. In both tests, the activity of daunomycin and adriamycin was found to be a linear function of the dose. However, as the dose-effect curve of foci inhibition has a steeper slope than that of cell proliferation, the two lines cross over. Hence higher doses give greater inhibition of foci formation than of cell proliferation. Almost total inhibition of foci formation could be achieved by treatment with daunomycin or adriamycin at about 0.025 μg/ml.

Dihydro daunomycin caused a remarkable reduction of foci formation, complete inhibition being obtained at 0.1 μg/ml. Foci formation was more sensitive to this compound than cell proliferation. N-guanidine daunomycin had no cytotoxic activity at the doses tested, while weakly inhibiting foci production.

The inhibitory activity of daunomycin and its structural analogues on viral oncogenesis by FLV and RSV, and on *"in vitro"* transformation by MSV (M) suggests that it is the activity of the virus-associated enzymes which is sensitive to these antibiotics. The RNA-dependent DNA polymerase of the virions is responsible for the synthesis of viral DNA. Table 16 shows how the reverse-transcriptase activity of MSV (M), FLV and RSV is inhibited by various daunomycin derivatives.

Table 16. Inhibition of reverse-transcriptase activity of RNA tumor viruses by daunomycin derivatives. Figures in brackets are percentages

System	Antibiotic (concentration = 100 μg/reaction mixture (0.25 ml))	^{3}H−TMP incorportaion into DNA (cpm/reaction mixture)		
		MSV (Moloney)	FLV	RSV
Without virions	–	7 (3.4)	7 (3.7)	7 (2.9)
Virions + RNase[1])	–	26 (13)	25 (13.4)	40 (16.8)
Complete	None	202 (100)	187 (100)	237 (100)
	Daunomycin	65 (32.1)	68 (36.3)	80 (33.7)
	Adriamycin	66 (33.1)	83 (44.4)	86 (36.3)
	Dihydro daunomycin	86 (42.5)	87 (46.5)	113 (47.7)
	N-guanidine daunomycin	106 (52.4)	97 (51.8)	117 (49.3)
	N-acetyl daunomycin	196 (97)	192 (102.6)	136 (57.3)

[1]) Virions containing Nonidet P-40 were preincubated at room temp. for 25 min with 50 μg/ml of pancreatic RNase.

Daunomycin and adriamycin at 10 μg/reaction mixture (0.25 ml) inhibit the reverse-transcriptase reaction by 60–70%. The dihydro derivative is also quite effective whereas the N-guanidine derivative has moderate activity. The N-acetyl derivative was completely ineffective in the MSV (M) and FLV systems but moderately inhibited the RSV system.

Table 17 shows how template-dependent DNA polymerase activity of FLV is inhibited by various daunomycin derivatives. The reactions catalyzed by poly-(dA-dT) and poly (rA) $\cdot$ (dT)$_{12}$ are highly sensitive to the action of daunomycin and its derivatives. Here again, daunomycin and adriamycin are most effective, and the N-acetyl derivative is completely inactive. It is interesting that the poly (dA-dT)-and poly (rA) $\cdot$ (dT)$_{12}$-dependent reactions are more sensitive to these antibiotics than the endogenous reaction (see Table 17), while the DNA polymerase reaction catalyzed by poly (dI-dC) is completely insensitive. The most active derivatives (daunomycin, adriamycin and dihydro daunomycin) slightly stimulate ^{3}H–dGMP incorporation catalyzed by poly (dI-dC). This stimulation is particularly noticeable in the case of dihydro daunomycin. Surprisingly, the N-acetyl derivative was found to inhibit this reaction. The mechanism of this inhibition is not understood.

Table 17. Inhibition of DNA-polymerase activity of FL virions by daunomycin and its derivatives in the presence of various templates. Figures in brackets are percentages

Antibiotic (5 μg/reaction mixture (0.25 ml))	^{3}H–TMP incorporation[1]) into DNA poly (dA-dT)	^{3}H–dGMP incorporation[1]) into DNA poly (dI-dC)	^{3}H–TMP incorporation[1]) into DNA poly rA $\cdot$ (dT)$_{12}$
None	1223 (100)	1006 (100)	723 (100)
Daunomycin	127 (10.3)	1057 (105)	159 (22)
Adriamycin	106 (8.7)	1127 (112)	231 (31.9)
Dihydro daunomycin	151 (12.3)	1654 (164.2)	327 (45.2)
N-guanidine daunomycin	322 (26.3)	941 (93.5)	457 (63.2)
N-acetyl daunomycin	1412 (115.6)	587 (58.3)	673 (93)

[1]) cpm/reaction mixture.

The inhibition of template-dependent DNA polymerase activity of MSV (M) by various daunomycin derivatives was also studied. Like the FLV system, the MSV poly (dA-dT)- and poly (rA) $\cdot$ (dT)$_{12}$-dependent reactions were extremely sensitive to daunomycin derivatives. In both cases the N-acetyl derivative was completely ineffective. Again as with the FLV system, we found that the poly (dI-dC) – catalyzed incorporation of ^{3}H–dGMP was not inhibited by any of the derivatives. In this system, unlike the FLV system, the N-acetyl derivative did not inhibit ^{3}H–dGMP incorporation into DNA. Dihydro daunomycin here too greatly stimulated ^{3}H–dGMP incorporation in the presence of poly (dI-dC).

The results show that the inhibition exerted by daunomycin derivatives against DNA polymerase from RNA tumor viruses is selectively dependent on the type of primer template used in the assay system. The inhibiting activity of daunomycin requires specific structural parameters. Thus, substitutions in the amino sugar moiety, especially N-acetylation, inhibit its activity against oncogenic viruses and influence its inhibitory action on the DNA polymerases of RNA tumor viruses.

To avaluate the therapeutic efficacy of these compounds, the activity of daunomycin and adriamycin on DNA-polymerases from various sources was measured. These studies were carried out using a constant concentration of the

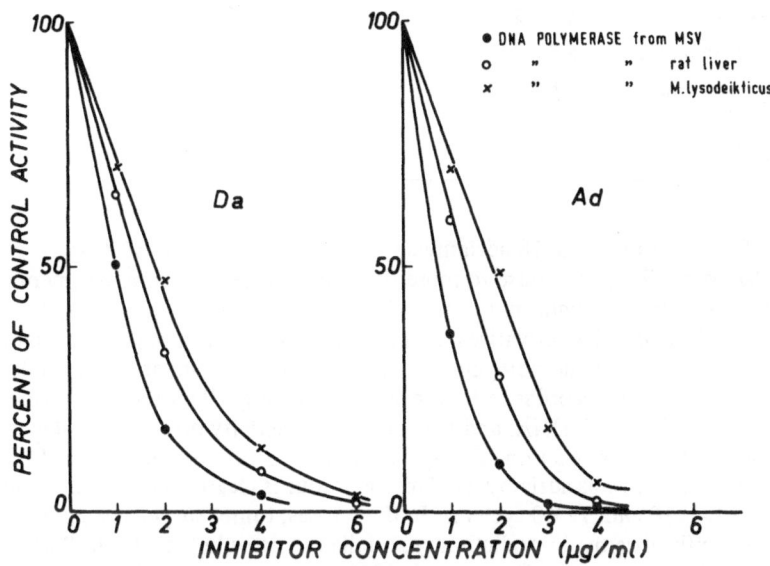

Fig. 15. Inhibition of poly (dA-dT)-directed activity of DNA polymerases from MSV (M), rat liver and *M. lysodeikticus*

template poly (dA-dT) in DNA-polymerase reactions, catalyzed by preparations from MSV (M), rat liver and *M. lysodeikticus*. As follows from results in Fig. 15, the MSV–DNA-polymerase is most sensitive to both the antibiotics.

The antitumor activity of some derivatives of daunomycin at the amino and methyl ketone functions has been studied by Yamamoto et al.[43]. Their studies were carried out mainly on leukemia 1210 in mice. At 2 mg/kg dose, the N-piperidinoimine derivative was found to have the same antitumor activity as daunomycin; other derivatives were not active at this dose level. The N-acetyl derivative was found to posses only a little antitumor activity, but displayed no accute toxicity even at very high doses. According to our experience, the N-acetyl derivative was in most of the cases ineffective against tumor growth (Table 18).

Table 18. Inhibition of growth of some transplanted tumors by daunomycin derivatives.
Tumor suspensions were in-cubated with 50 μg/ml of the antibiotic at 37 °C for 1 h

Antibiotic	10 animals transplanted with		
	Ridgeway-Osteo-Sarcoma (Mouse) Tumor weight (g)	L-1210 (Mouse) Ascites (ml)	SV 40 (Hamster) Tumor weight (g)
None	7.0 ± 2.4	0.78 ± 0.16	47 ± 12
Adriamycin	1.0 ± 0.3	0.00	0.0
Daunomycin	5.0 ± 2.3	0.00	0.0
Dihydro daunomycin	5.3 ± 2.0	0.53 ± 0.2	0.0
N-guanidine-acetamide daunomycin	7.8 ± 2.8	0.63 ± 0.1	18.0 ± 8
N-acetyl daunomycin	9.8 ± 1.4	0.64 ± 0.07	46 ± 8

As follows from Table 18 adriamycin inhibits the growth of *Ridgeway-Osteo-Sarcoma* (ROS) in mouse to more than 80%. Under similar experimental conditions one finds a slight inhibitory effect by daunomycin and its dihydro derivative. However, the derivatives with substitutions in the aminosugar moiety, N-gaunidino-acetamide-daunomycin and N-acetyl-daunomycin, are completely ineffective. This is in accordance to our previous findings[27] on the interaction of these derivatives with DNA, and their inhibitory activity on the DNA-dependent RNA polymerase reaction. Though, in these studies daunomycin was found to be almost as active as adriamycin. This behaviour is clearly demonstrated on studies in L-1210 and SV 40 systems. In these cases, tumor suspensions pre-incubated with adriamycin or daunomycin failed to grow in their hosts. The dihydro derivative was not effective in L-1210, however a total inhibition was achieved in case of SV 40. Unexpectedly, the N-guanidino-acetamide derivative showed a significant activity against SV 40. However, the N-acetyl derivative was ineffective against all types of tumors studied by us.

3. Tilorone Hydrochloride

The dihydro-chloride salt of 2,7-bis(2-(diethylamino)ethoxy)-fluoren-9-one, referred to as tilorone hydrochloride (non-proprietary name) or bis-DEAE-fluorenone, is a broad spectrum antiviral compound[44] with antitumor activity[45-47]. Mayer and coworkers[48, 49] have identified this compound as an interferon inducer and established a relationship with the antiviral activity. However, recently a lack of correlation between interferon induction and viral protection by tilorone hydrochloride has been reported[50].

The possibility that this compound may react directly with DNA was indicated by the cytogenetic studies of Green and West[51]. Tilorone was found to

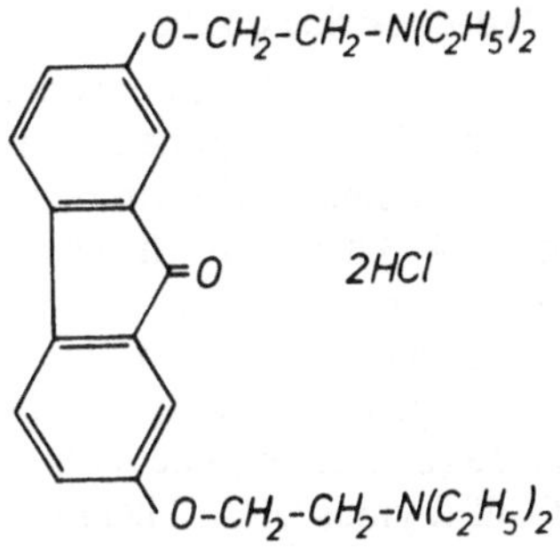

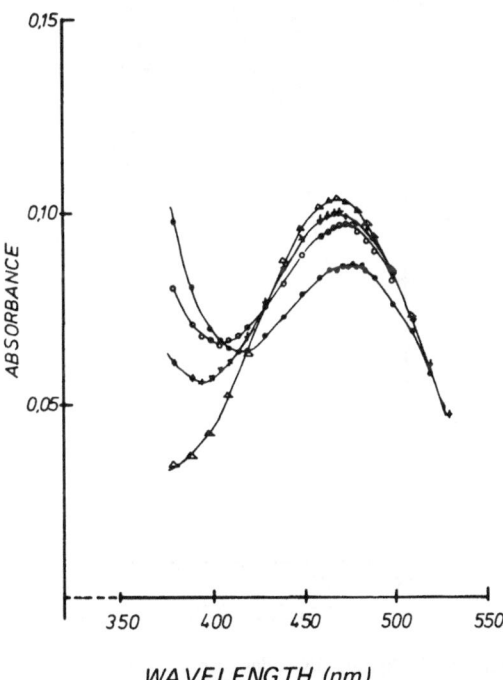

O-CH$_2$-CH$_2$-N(C$_2$H$_5$)$_2$

=O   2HCl

O-CH$_2$-CH$_2$-N(C$_2$H$_5$)$_2$

Fig. 16. Chemical structure of tilorone hydrochloride

inhibit mitosis significantly at 3.0 μg/ml, and produced chromosomal abnormalities at 1.5 μg/ml. Soon it was discovered by Chandra *et al.*[52, 53a, 53b] that tilorone does form molecular complexes with DNA and poly-deoxynucleosides. Some of these studies will be described here.

Fig. 17. Effect of calf thymus DNA on the visible absorption spectrum of tilorone hydrochloride. Samples contained 4,25 x 10^{-4}M of tilorone hydrochloride, 0.01 M Tris-HCl (pH 7.0) and DNA at 0.5 x 10^{-3}M (+ ——— +); 1 x 10^{-3}M (0 ——— 0); 2 x 10^{-3}M (0 ——— 0). No DNA was added to the sample (Δ ——— Δ)

125

3.1. Influence of Tilorone Hydrochloride on the Secondary Structure of DNA

Interaction between nucleic acids and biologically active compounds may induce changes in the electronic spectra of the components. Tilorone hydrochloride in water shows two absorption maxima, in the ultraviolet region around 271 nm, and in the visible region around 470 nm. Thus the investigation of the long wavelength band, where DNA and RNA do not absorb, should provide some evidence whether or not the chromophore of tilorone hydrochloride is involved in the binding process. Fig. 17 depicts the *absorption spectrum* (350–500 nm) of tilorone hydrochloride alone (continuous line with triangles) or in the presence of various amounts of calf thymus DNA. There is a characteristic change

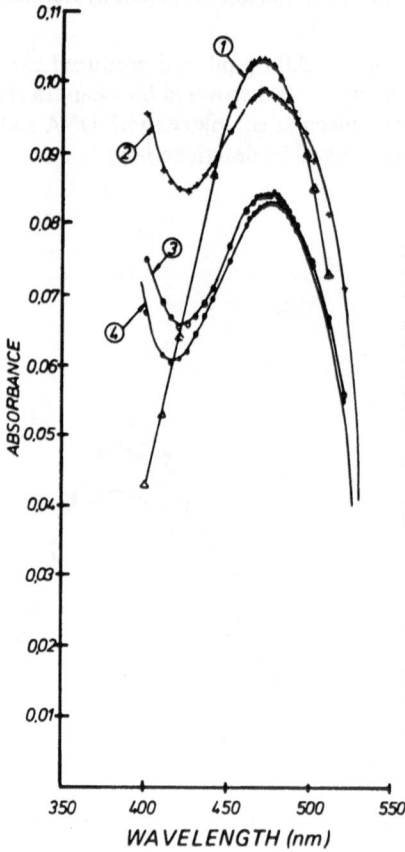

Fig. 18. Effect of native calf thymus DNA, denatured calf thymus DNA and yeast RNA on the visible absorption spectrum of tilorone in 0.01 M Tris-HCl (pH 7.0). Curve *1* is the spectrum of free tilorone (4,25 x 10^{-4}M). Other curves depict the spectra of tilorone in the presence of yeast RNA (curve *2*), denatured DNA (curve *3*) and native DNA (curve *4*). Molar concentrations of nucleic acids (2x10^{-3}M) refer to phosphorus content of the polymer

in tilorone spectrum in the presence of DNA. In the presence of calf thymus DNA the visible absorption spectrum of tilorone hydrochloride is depressed and red shifted. This hypochromic effect of DNA on the absorption of tilorone chromophore is dependent on DNA concentration. The largest hypochromic effect is observed at 2×10^{-3} M DNA-P in a $4{,}25 \times 10^{-4}$ M solution of tilorone hydrochloride.

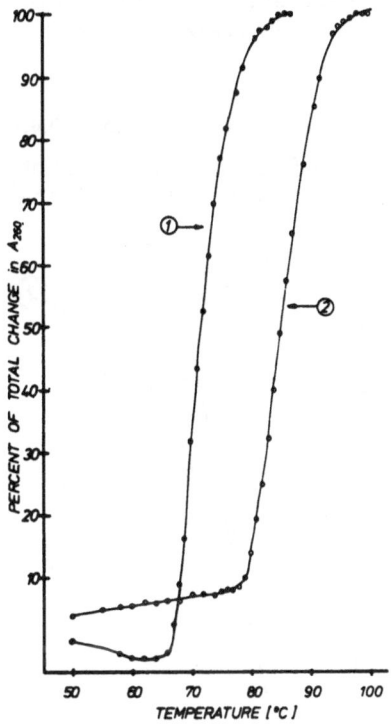

Fig. 19. Effect of tilorone on the thermal transition temperature (T_m) of calf thyms DNA. Solvent is 0.01 M Tris. HCl (pH 7.0), and the DNA concentration is 5×10^{-5} M in all experiments. Curve *1* represents the melting profile of DNA in the absence of tilorone, and curve *2* is the melting profile of DNA in the presence of 1×10^{-5} M tilorone hydrochloride

The concentration-dependent effect of calf thymus DNA on the visible absorption spectrum of tilorone hydrochloride indicates that the tilorone chromophore interacts with DNA. Fig. 18 depicts the visible absorption spectra of tilorone alone (curve *1*), or in the presence of yeast RNA (curve *2*), denatured DNA (curve *3*) and native double-stranded DNA (curve *4*). The visible spectra indicate that at equimolar concentrations, DNA in its double helical state produces largest changes in the absorption spectrum of tilorone, whereas the effect of single-stranded DNA is slightly weaker. In contrast, the yeast RNA exerts only a slight effect on the visible spectrum of tilorone hydro-

127

chloride. These data indicate a specificity of the tilorone chromophore towards DNA.

Further information on the binding of tilorone with DNA was derived by studying the *thermal melting* of the complex. In order to characterize the stability of DNA secondary structure in the presence of tilorone, temperature profiles were run at tilorone/DNA-P molar ratio of 1 : 5 (Fig. 19). Tilorone hydrochloride shows a large increase in the thermal transition temperature (T_m) of native DNA; the T_m of calf thymus DNA was raised from 71.6 to 85.2 °C under these conditions.

3.2. Mode of Tilorone Hydrochloride Interaction to DNA

Hypochromic effect of native DNA on the absorption of tilorone chromophore is partially reversible by Na^+ and Mg^{2+} ions. Fig. 20 depicts the absorption spec-

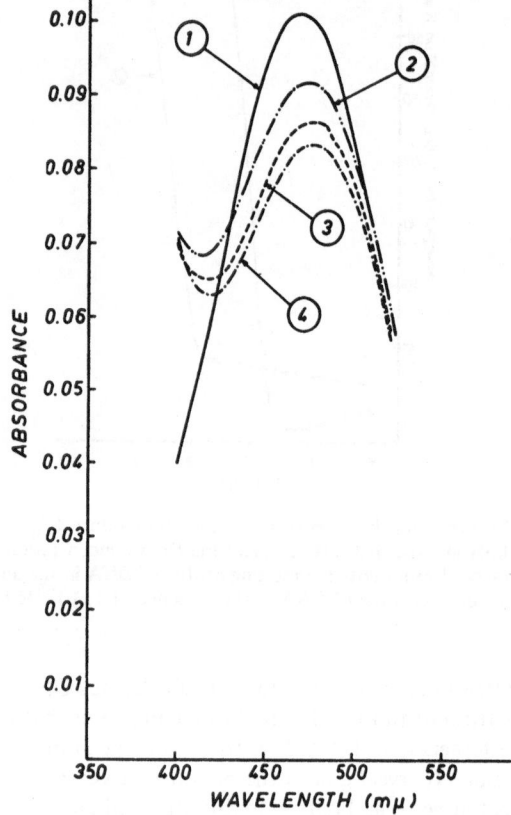

Fig. 20. Effect of Na^+ and Mg^{2+} on the visible absorption spectrum of the tilorone-DNA complex. Samples contained $4,25 \times 10^{-4}$M tilorone, 4×10^{-3}M DNA-P, 0.01 M Tris-HCl (pH 7.0) and 0.01 M $MgCl_2$ (curve 2) or 0.1 M NaCl (curve 3). Curve 1 is the spectrum of free tilorone; curve 4 is the spectrum of the tilorone-DNA complex in the absence of Na^+ and Mg^{2+}

tra (350–550 nm) of tilorone hydrochloride alone, 4,25 x 10^{-4} (curve *1*), in the presence of 4 x 10^{-3} M DNA-P (curve *4*) containing 0.01 M $MgCl_2$ (curve *2*) or 0.1 M NaCl (curve *3*). It follows from these results that the DNA-drug interaction is very sensitive to magnesium ions. The effect of magnesium ions on tilorone binding to DNA was confirmed by density-gradient studies using labeled tilorone hydrochloride.

These studies indicate that electrostatic forces contribute greatly to the binding process. The interaction between tilorone and DNA may, however, involve other kinds of forces. Tilorone forms a reversible complex with DNA, since the drug could be completely dissociated from a DNA-cellulose column. Interaction of apurinic and apyrimidinic DNA's with tilorone hydrochloride also gave spectral changes. However, only with the apyrimidinic DNA, the spectrum of the bound drug was similar to that found with native DNA.

The *absorption spectrum* studies presented above merely reflect the electronic environment of the molecule and do not give specific information about the type of interaction. The data which must be accounted for in considering a physical mode for the binding process can be derived from several different approaches. Hydrodynamic measurements on the DNA-drug complex are of interest, since Lerman[42] has established that an increase in the intrinsic viscosity of DNA and a decrease in the sedimentation coefficient of the polymer are two criteria for intercalation of ring systems between base pairs of a double-helical DNA.

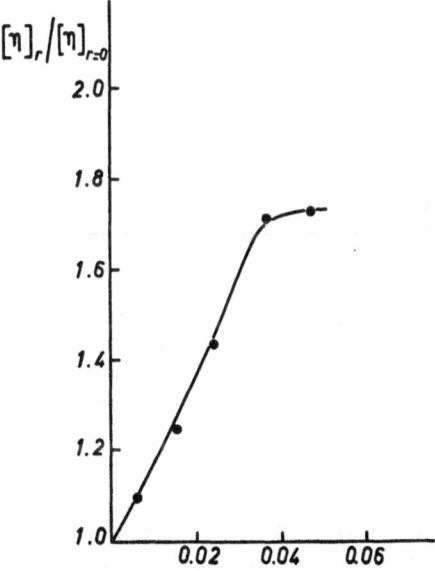

Fig. 21. Intrinsic viscosity of tilorone-DNA complex relative to intrinsic viscosity of DNA alone. Conditions: 20 °C, 0,01 M Tris-HCl, pH 7.0

Fig. 21 shows the relationship between the *intrinsic viscocity* of DNA and the amount ("*r*") of bound tilorone. The intrinsic viscosity of the complex increases with *r* up to a limiting value of about 0.05. The maximum relative enhancement of viscosity was about

1.7. In addition, at the same ionic strength and at a ligand to DNA-P molar ratio of 0.1, the sedimentation rate of DNA was decreased to 78% of the value in the absence of ligand.

These observations are consistent with an intercalative mode of binding in the interaction of tilorone hydrochloride with double-helical DNA. These results were not examined in attempt to verify whether they agree with measurements of the length increase on sonicated DNA. For this reason, the intercalation model of the DNA complex remains tentative.

The interaction of tilorone hydrochloride with native DNA *stabilizes the double helical structure* of the macromolecule towards thermal denaturation. The effect of tilorone hydrochloride on the thermal denaturation of DNA's from various sources having different base composition is shown in Fig. 22.

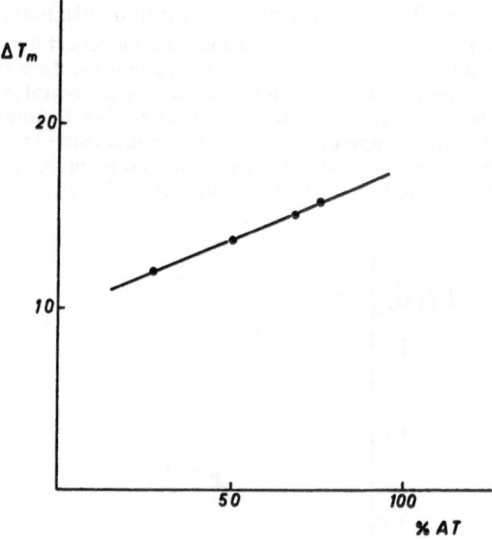

Fig. 22. Effect of tilorone on the T_m with DNA preparations from various sources of different base compositions. T_m values were obtained at a molar ratio of drug to DNA-P of 0,21. The T_m of DNA (5×10^{-5}M) with and without tilorone was determined in a medium of 0.01 M Tris-HCl (pH 7.0). AT-content of *B. cereus* DNA (65%), calf thymus DNA (58%), *E. coli* (DNA (50%) and *M. lysodeikticus* DNA (28%); these DNA samples were used in these experiments

At a drug to DNA-P molar ratio of 0,21, the ΔT_m increased with increasing AT content of the DNA. This observation indicates that tilorone hydrochloride preferentially binds to the dAT portions of the DNA molecule. This hypothesis is confirmed by the strong effect of tilorone hydrochloride on the thermal transition temperature of poly d(A-T), ΔT_m = 29 °C.

An intercalative mechanism for binding of a ligand to DNA is consistent with a stabilization of the double helix. Such as stabilization, however, does not constitute proof of intercalation. But, when considered with the evidence of the results reported in this paper, showing increased viscosities and decreased sedimentation rate of DNA, one may conclude that the large increase of T_m points to an intercalative mode of binding.

3.3. Effect of Tilorone Hydrochloride on the Template Activity of Nucleic Acids

The interaction of tilorone hydrochloride to native as well as denatured DNA encouraged us to study the template activity of the complexes in DNA- and RNA-polymerase systems from *E. coli*. Table 19 shows the effect of tilorone on the priming activity of denatured DNA in DNA-polymerase reaction. DNA-polymerase was isolated from *E. coli* B cells according to the procedure of Richardson[54], and fraction VII obtained after DEAE-cellulose chromatography was used. The reaction was carried out in the presence of denatured DNA.

Table 19. Inhibition of DNA-dependent DNA polymerase reaction (*E. coli* B) by tilorone hydrochloride. The calf thymus DNA primed assay system contained (total vol 0.3 ml) 0.07 M glycine buffer, pH 9.2, 7 mM MgCl$_2$, 1 mM β-mercaptoethanol, 10 mμmoles each of dTTP, dCTP and dGTP, 2 μCi of [3]H−dATP, 20 μg of denatured calf thymus DNA. The reaction was started by adding 0.02 ml (approx. 50 μg protein) of the enzyme preparation

System	Tilorone hydrochloride concentration (μg/0.30 ml reaction mixture)	[3]H−dAMP incorporation into DNA (cpm/reaction mixture)	% of Control
Without DNA	−	67	0.2
Without enzyme	−	35	0.1
Complete	None	29,087	100
	5	5,395	18.3
	10	1,380	4.7
	15	372	1.2
	20	166	0.57

As follows from the results there is a concentration dependent inhibition of ^{3}H-dAMP incorporation into DNA by tilorone hydrochloride. Concentration as low as 5 μg per reaction mixture inhibits more than 80% the incorporation of [3]H−dAMP into DNA. At 15 μg/reaction mixture the reaction is completely blocked by tilorone.

Table 20. Inhibition of DNA-dependent RNA polymerase reaction (*E. coli* K-12) by tilorone hydrochloride

System	Tilorone hydrochloride concentration (μg/0.25 ml reaction mixture)	^{14}C–AMP incorporation into RNA (cpm/reaction mixture)	% of control
Without UTP, CTP and GTP	–	14	2.1
Without DNA	–	11	1.6
Complete	None	652	100
	25	471	72
	50	313	46
	100	201	31

The inhibiting activity of tilorone hydrochloride on the DNA-dependent RNA-polymerase reaction is shown in Table 20. Compared to the DNA-polymerase reaction, the RNA-polymerase reaction requires large amounts of tilorone hydrochloride for its inhibition; no significant inhibition was observed below 15 μg/reaction mixture of tilorone hydrochloride. Whereas this amount of tilorone was able to completely inhibit the DNA polymerase reaction (see Table 19). One explanation is that in the RNA-polymerase reaction the DNA concentration is approx. 2.5 times more than that used in the DNA-polymerase reaction. However, this may not be the only reason for such differences. Our spectrophotometric data show that Mg^{2+}ions influence the tilorone binding to DNA. Since the Mg^{2+}ion concentrations in both systems are different, this may account for the variable sensitivity of both systems towards tilorone hydrochloride.

Table 21. Inhibition of reverse-transcriptase activity of RNA tumor viruses by tilorone hydrochloride. Figures in parentheses are percent of control

System	Tilorone hydrochloride concentration (μg/reaction mixture (0.25 ml))	^{3}H–TMP incorporation into DNA (cpm/reaction mixture)	
		MSV (Moloney)	FLV (Friend)
Without virions	–	7 (3.2)	7 (2.6)
Virions + RNase[1]	–	31 (14.3)	39 (14.8)
Complete	None	216 (100.0)	262 (100.0)
	5	112 (51.8)	189 (72.0)
	10	88 (40.5)	128 (48.8)
	20	57 (26.4)	81 (30.9)

[1]) Virions containing Nonidet P-40 were preincubated at room temp. 25 min with 50 μg/ml of pancreatic RNase.

Munson *et al.*[47)] have recently shown that DEAE-F is effective in inhibiting the established *Friend viral leukemia*. They believe that interferon induction may not be responsible for the antitumor activity of this compound. This suggests that the virus-associated enzymatic activities, DNA polymerases, may be sensitive towards the action of DEAE-F. Table 21 shows the inhibition of reverse-transcriptase activity from MSV (Moloney) and FLV (Friend) by DEAE-F.

Results presented in Table 21 show that the in vitro system is dependent on the source of enzyme, *i.e.* virions, and sensitive to RNase. Preincubation of virions with RNase blocks their activity to synthesize DNA. This shows that the endogenic template, viral RNA, is required for the synthesis of DNA. DEAE-F added to the reaction mixture inhibits the DNA-polymerase activity in MSV (M) and FLV. At low concentrations (5 μg/reaction mixt.) of DEAE-F the MSV (M) system is more sensitive than FLV. However, at higher concentrations of DEAE-F the inhibition in both the systems is of the same magnitude. It is interesting to note that concentrations as low as 20 μg/reaction mixt. are able to inhibit approximately 70% of incorporation of ^{3}H–TMP into DNA.

Table 22. Inhibition of DNA-polymerase activity from FLV (Friend) by tilorone hydrochloride in the presence of various templates. The figures in parentheses indicate the percent of control (without DEAE-F)

Tilorone hydrochloride concentration (μg/0.25 ml reaction mixture)	^{3}H–TMP incorporation into DNA (cpm/reaction mixture)			^{3}H–dGMP incorporation into DNA (cpm/reaction mixture)
	Poly (dA-dT)	Poly (rA · dT)	Poly rA · (dT)$_{12}$	poly (dI-dC)
None	2387 (100.0)	5330 (100.0)	572 (100.0)	1797 (100.0)
5	460 (19.2)	5170 (97.0)	493 (86.2)	3707 (206.3)
10	292 (12.2)	3619 (67.9)	317 (55.4)	3962 (220.4)
20	190 (8.0)	2808 (52.6)	212 (36.5)	6842 (380.7)

Synthetic polymers containing either desoxyribonucleotide or ribonucleotide strands are known to stimulate the in vitro DNA synthesis by RNA tumor viruses. Some inhibitors of the DNA-polymerase reaction in RNA tumor viruses are known to exhibit a template-primer specificity[6, 7, 34)]. Table 22 shows the inhibition of template-dependent DNA polymerase activity of FLV by DEAE-F at various concentrations. The reaction catalyzed by poly (dI · dC) is most strongly stimulated by DEAE-F. Thus at 20 μg/reaction mixt. of DEAE-F the incorporation of ^{3}H–dGMP is almost 4 times that of control.

The present results show that the inhibition exerted by DEAE-F against DNA polymerases from RNA tumor viruses is uniquely and selectively dependent on the type of primer-template used in the assay system. The wide differences between the inhibitory concentrations of DNA-dependent enzymatic

reactions in RNA tumor viruses and bacteria could be of a significant therapeutic value. However, it would require more studies on the DNA-dependent
reactions in animal cells before one may speculate on its therapeutic superio-
rity. Studies on these aspects are in progress in our laboratory.

4. Modified Nucleic Acids

Mercapto-(5-SH)-polycytidylic acid has been shown by Bardos *et al.*[55] to
block the DNA-directed RNA synthesis at 1/2 to 1/50 of the concentration
of the unmodified DNA template used in the reaction. This mercapto derivative of polycytidylic acid (MPC) has been prepared by partial "thiolation" of
polycytidylic acid (PC) according to the general procedure of Bardos *et al.*[56, 57];
the thiolated compound was gel-filtered through a Sephadex column and subsequently, through an Agarose- 1.5 m (Bio-Gel A-1.5 m, exclusion limit
1,5000.000 mol. wt., Bio-Rad Labs.) column, then lyophilized and redissolved
in 0.1 M Tris-buffer[58]. This compound (conversion of 9.5% of the cytidylate
units to 5-mercaptocytidylate) was studied in the DNA-polymerase system of
oncorna viruses.

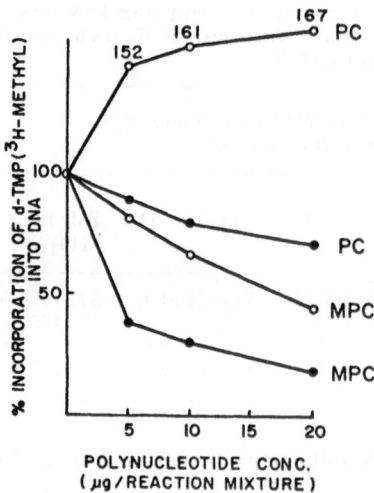

Fig. 23. Effects of thiolated polycytidylic acid (MPC) and unmodified polycytidylic acid
(PC) on the DNA polymerase activities from MSV (Moloney) in the presence of poly rA·
(dT)$_{14}$ (●——●) and poly (dA-dT) (○——○) as templates

Fig. 23 shows the effects of 5-mercapto-(9.5%)-polycytidylic acid (MPC)
and modified polycytidylic acid (PC), respectively, on the incorporation of
^{3}H–TMP into DNA by the DNA-polymerses of the MSV-M, in the presence of
either poly(dA-dT) or poly rA · (dT)$_{14}$ as the template. The results obtained
with the same pair of modified and unmodified polycytidylic acid samples in

the FLV DNA-polymerase assay, using the same pair of templates, are graphically represented in Fig. 24

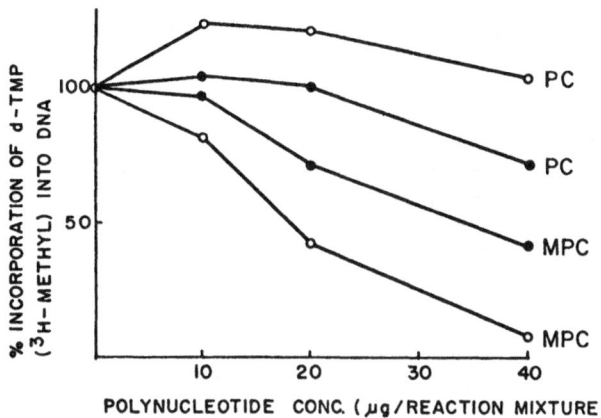

Fig. 24. Effects of thiolated polycytidylic acid (MPC) and unmodified polycytidylic acid (PC) on the DNA polymerase activities from FLV (Friend) in the presence of poly rA· $(dT)_{14}$ (●——●) and poly (dA-dT) (○——○) as templates

It is clear from these graphs that the modified polynucleotide, MPC, significantly inhibits the DNA polymerases present in both viral extracts; furthermore, the inhibitory activity of MPC is very nearly the same in the two systems when poly (dA-dT) is used as the template (50% inhibition at 18 μg/r.mix.), but in the presence of poly rA · $(dT)_{14}$ as the template, MPC acts as a much more potent inhibitor of ^3H—TMP incorporation in the MSV-M assay system (50% inhibition at 4 μg/r.mix.) than in the FLV system (50% inhibition at 35 μg/r. mix.). In contrast, the unmodified polynucleotide, PC, stimulates DNA polymerase activity in both viral systems with poly (dA-dT) as the template, and it shows slight inhibitory activity (only 25% inhibition, at 20–40 μg/r.mix.) in the presence of the poly rA · $(dT)_{14}$ template.

It is of considerable interest that this thiolated polynucleotide is capable of distinguishing between two viral "reverse transcriptases" and that in the case of the MSV-M enzymes, it shows much greater inhibition of the "reverse transcriptase" than of the DNA-directed DNA polymerase. This suggests the possibility that other thiolated polynucleotides, more closely resembling the RNA of a specific virus than MPC, may show even greater selectivities as inhibitors of the corresponding reverse transcriptase. Such modified polynucleotides are currently being prepared and tested, in the hope that they may prove to be useful tools for the identification and study of the reverse transcriptases found in normal and malignant tissues and of the possible relationships of the latter to oncogenic viruses. In addition, the feasibility of using such modified polynucleotides as selective inhibitors in cancer chemotherapy is being examined.

135

Recently we have thiolated *nucleic acid fractions isolated from Ehrlich ascites (EA) cells*. The activities of thiolated DNA, ribosomal RNA and transfer RNA were studied in a DNA-polymerase system from Friend leukemia virions. Of all the fractions tested transfer RNA, thiolated 1−3% showed a maximum inhibition of DNA polymerases of oncorna viruses. Table 23 shows some of the results obtained using thiolated tRNA.

Table 23. Effect of EA-tRNA on DNA polymerases of FL-virions

System	% Incorporation of Control		
	Endogenic Reaction	Poly rA · $(dT)_{14}$	Poly d(A-T)
Non-thiolated (μg/reaction mixture)			
10	91	115	97
20	87	121	96
40	85	126	81
Thiolated (1−3%)			
10	50	34	22
20	45	22	15
40	36	16	12

As follows from Table 23 the unmodified EA-tRNA does not inhibit the DNA polymerase activity of FLV at the concentrations used in the incubation mixture. However, at the same concentrations the thiolated EA-tRNA is a very strong inhibitor of DNA-polymerase activities, endogenous as well as template-dependent, of FLV. Experiments are in progress to modify nucleic acid fractions of viral origin.

5. Conclusion and Future Prospects

Soon after the discovery of DNA polymerases in virions of RNA tumor viruses, a great deal of hope was expressed that the discovery might lead to resolve the possibility of the involvement of oncorna viruses in an inapparent form in "spontaneous" or chemically induced tumors, especially in man. So far, there has been some evidence in support that RNA tumor viruses are related to human neoplasia.

Gallo and his colleagues[3] were the first to report some kind of RNA-directed DNA polymerase in tumor cells, but not in normal cells. They were able to purify from the leukemic cells, but not the normal cells, a DNA polymerase which would use both natural RNA's and poly A · poly (dT) as a template. Moore et al.[59] reported small numbers of particles resembling virions in human milk. These particles were found in high frequency in the milk of Parsi (Indian) women and in relatives of people with breast cancer. These particles banded at a density of $1.16 - 1.19$ gm/cm^3 and were shown by Schlom et al.[60] to contain a ribonuclease-sensitive DNA polymerase activity. Hirschman et al.[61] reported a DNA polymerase in some partially purified preparations of Australia antigen. However, there was no correlation of polymerase activity and amount of antigen in different sera. Gallo et al.[62] have reported DNA polymerase activity in the particles, ESP-1, originally described by Priori et al.[63] in the supernatant of a human cell line. The particles banded at a density of 1.15 gm/cm^3.

The presence of enzymes which catalyze RNA-directed DNA synthesis in virus-like particles, or neoplastic cells of human origin, described above, do not necessarily confirm the role of RNA viruses in human neoplasia. However, these discoveries lead to a possibility that studies with compounds, which are potent and specific inhibitors of such enzyme(s) may be useful to design antitumor compounds of future interest.

Acknowledgement

The author is much indebted to Doctors F. Zunino, A. M. Casazza, D. Gericke and T. J. Bardos for collaboration in several o f the experiments reported in this article. Gratefully acknowledged are Mrs. A. Götz und Mr. A. Zaccara for their expert technical assistance and Mrs. A. Woelke for preparing the manuscript.

P. Chandra

References

1) Temin, H. M., Mitzutani, S.: Nature 226, 1211 (1970).
2) Baltimore, D.: Nature 226, 1209 (1970).
3) Gallo, R. C., Yang, S. S., Ting, R. C.: Nature 228, 927 (1970).
4) Chandra, P., et al.: FEBS-Letters 16, 249 (1971).
5) Chandra, P., et al.: FEBS-Letters 19, 327 (1972).
6) Chandra, P., et al.: FEBS-Letters 21, 154 (1972).
7) Chandra, P., et al.: FEBS-Letters 21, 264 (1972).
8) Gallo, R. C., et al. in: The biology of oncogenic viruses (ed. L.S. Silvestri), p. 210. Amsterdam: North. Holland Publ. Co. 1971.
9) Green, M., et al. in: The biology of oncogenic viruses (ed. L. S. Silvestri), p. 193. Amsterdam: North-Holland Publ. Co. 1971.
10) Kotler, M., Becker, Y.: Nature New Biol. 234, 212 (1971).
11) Arcamone, F., in: IUPAC (Paris 1957), Resume des Comm. 2, 194 (1957).
12) Di Marco, A., et al.: Cancer Chemother. Repts. 18, 15 (1962).
13) Di Marco, A., Soldati, M., Fioretti, A.: Acta Un. Int. Cancer 20, 423 (1964).
14) Arcamone, F., et al.: Nature 203, 1064 (1964).
15) Finlay, A. C., et al.: J. Am. Chem. Soc. 73, 341 (1951).
16) Julia, M. M., Preau-Joseph, N.: Compt. Rend. 257, 1115 (1963).
17) Schabel, F. M., et al.: Proc. Soc. Exptl. Biol. Med. 83, 1 (1953).
18) Werner, G. H., Maral, R.: Actualite's Pharmaceutiques 21, 156 (1963).
19) Zunino, F., Di Marco, A.: Biochem. Pharmacol. 21, 867 (1972).
20) Zimmer, Ch., et al.: J. Molec. Biol. 58, 329 (1971).
21) Krey, A. K., Hahn, F.: FEBS-Letters 10, 175 (1970).
22) Chandra, P., Zimmer, Ch., Thrum, H.: FEBS-Letters 7, 90 (1970).
23) Chandra, P., et al.: Hoppe-Seyler's Z. Physiol. Chem. 353, 393 (1972).
24) Inagaki, A., Kageyama, M.: J. Biochem. 68, 187 (1970).
25) Arcamone, F., et al.: Gazetta. Chim. Ital. 99, 620 (1969).
26) Arcamone, F., et al.: Gazetta Chim. Ital. 99, 632 (1969).
27) Chandra, P., et al.: Naturwissenschaften 59, 448 (1972).
28) Zimmer, Ch., Luck, G., Thrum, H.: Stud. biophys. 24/25, 311 (1970).
29) Chandra, P., Götz, A., Hausmann, S.: to be published.
30) Herrmann, E. C., et al.: Proc. Soc. Exptl. Biol. Med. 103, 625 (1960).
31) Burgess, R. R.: J. Biol. Chem. 244, 6160 (1969).
32) Casazza, A. M.: VII. Intn. Conf. Chemother., Prague 1971, I, A-5/41.
33) Hirschmann, S. H., et al.: J. Natl. Cancer Inst. 42, 399 (1969).
34) Fridlender, B., Weissbach, A.: Proc. Natl. Acad. Sci., USA 68, 3116 (1971).
35) Grein, A., et al.: Microbiol. 11, 109 (1963).
36) Di Marco, A., et al.: Nature 201, 706 (1964).
37) Arcamone, F., et al.: J. Am. Chem. Soc. 86, 5334 (1964).
38) Arcamone, F., et al.: J. Am. Chem. Soc. 86, 5335 (1964).
39) Calendi, E., et al.: Biochim. Biophys. Acta 103, 25 (1965).
40) Di Marco, A. in: Antibiotics (eds. D. Gottlieb and P. D. Shaw), Vol. I, p. 190, Berlin Heidelberg–New York: Springer 1967.
41) Hartmann, G., et al.: Biochem. Z. 341, 126 (1964).
42) Lerman, L. S.: J. Molec. Biol. 3, 18 (1961); J. Cell. Comp. Physiol. 64, 1 (1964).
43) Yamamato, K., Acton, E. M., Henry, D. W.: J. Med. Chem. 15, 872 (1972).
44) Krüger, R. F., Yoshimura, S.: Federation Proc. 29, 635 (1970).
45) Adamson, R. H.: J. Natl. Cancer Inst. 46, 431 (1971).
46) Munson, A. E., Munson, J. A., Regelson, W.: Intl. Colloq. on Interferone inducers, Leuven, Belgium, Abstract No. 27 (1971).
47) Munson, A. E., et al.: Cancer Res. 32, 1397 (1972).
48) Mayer, G. D., Krüger, R. F.: Science 169, 1214 (1970).
49) Mayer, G. D., Fink, B. A.: Federation Proc. 29, 635 (1970).

138

50) Giron, D. J., Schmidt, J. P., Pindak, F. F.: Antimicrobial agents and Chemother. *1*, 78 (1972).
51) Green, S., West, W. L.: Pharmacologist, *13*, 260 (1971).
52) Chandra, P., Zunino, F.,Götz, A.: FEBS-Letters *22*, 161 (1972).
53) a) Chandra, P., *et al.*: FEBS-Letters *23*, 145 (1972).
    b) Chandra, P., *et al.*: FEBS-Letters *28*, 5 (1972).
54) Richardson, C. C. in: Procedures in nucleic acid research (eds. G. L. Cantoni and D. R. Davis) p. 263. London: Harper & Row 1966.
55) Bardos, T. J., *et al.*: Proc. 8th Intl. Symp. on the Chemistry of Natural Products, New Delhi, India, February 1972, p. 329.
56) Bardos, T. J., *et al.*: Proc. Amer. Assoc. for Cancer Res. *13*, 359 (1972).
57) Bardos, T. J., *et al.*: 163rd Meeting of Am. Chem. Soc., Abstract *21* (1972).
58) Chandra, P., Bardos, T. J.: Res. Commun. Chem. Pathol and Pharmacol *4*, 615 (1972).
59) Moore, D. H., *et al.*: Nature *229*, 611 (1971).
60) Schlom, J., Spiegelman, S., Moore, H.: Nature *231*, 97 (1971).
61) Hirschman, S. Z., Vernace, S. J., Schaffner, F.: Lancet *1*, 1099 (1971).
62) Gallo, R. C., *et al.*: Nature *232*, 140 (1971).
63) Priori, E. S., *et al.*: Nature New Biol. *232*, 61 (1971).

Received April 2, 1973

Alkylating Agents

Dr. Thomas A. Connors

Chester Beatty Research Institute, Institute of Cancer Research, Royal Cancer Hospital, London, England

Contents

Unique Biological Properties

The unique biological properties of at least two alkylating agents were first observed at the end of the last century [1-3], but their pharmacology was not studied intensively until some 40 years later when a series of alkylating agents, the aliphatic nitrogen mustards (di-(2-chloroethyl)alkylamines) were investigated because of their relationship to the chemical warfare agent, sulphur mustard gas (di-2-chloroethyl sulphide) *1*. Although sulphur mustard had been used earlier in attempts to treat cancer in animals and man [4, 5], it was the recognition by Gilman and his colleagues [6, 7] of the specific effects of an aliphatic nitrogen mustard on lymphoid tissue and rapidly dividing cells that first suggested that they might be used with advantage in the treatment of lymphomas and leukaemias. Early encouraging results with tris-(2-chloroethyl)-amine *2* and di-(2-chloroethyl)methylamine (HN2; *3*) led to the synthesis and testing of related derivatives[a] in attempts to obtain agents which maintained

$$S \underset{CH_2CH_2Cl}{\overset{CH_2CH_2Cl}{<}} \qquad Cl \cdot CH_2CH_2 \cdot N \underset{CH_2CH_2Cl}{\overset{CH_2CH_2Cl}{<}} \qquad CH_3 \cdot N \underset{CH_2CH_2Cl}{\overset{CH_2CH_2Cl}{<}}$$

1   *2*   *3*

their anti-tumour potency but had less undesirable side-effects seen with these early chemicals. It was soon recognised that while aromatic nitrogen mustards such as N,N-di-2-chloroethylaniline (aniline mustard; *4*), could be just as effective as the aliphatic congeners in their anti-tumour action, two alkylating arms were essential for activity. Monofunctional aromatic or aliphatic nitrogen mustards such as N-2-chloroethylaniline *5* and 2-chloroethyldimethylamine *6* had no effect on the growth of animal tumours.

$$\underset{4}{\text{⟨⟩—N} \underset{CH_2CH_2Cl}{\overset{CH_2CH_2Cl}{<}}} \qquad \underset{5}{\text{⟨⟩—NH·CH}_2CH_2Cl} \qquad \underset{6}{\overset{CH_3}{\underset{CH_3}{>}} N \cdot CH_2 \cdot CH_2 \cdot Cl}$$

Certain exceptions are known, but even where monofunctional nitrogen mustards are active they do not have the same broad spectrum of action and usually contain another functional group in the molecule [8-10].

Once it was appreciated that the nitrogen mustards were acting as electrophilic reactants, alkylating essential cellular macromolecules, other types of chemical with similar properties were tested for their tumour inhibitory action.

---

[a] H. Arnold, Topics Curr. Chem. 7, 196 (1963).

Useful compounds were found amongst the sulphonoxy alkanes such as 1,4-disulphonoxybutane (7; myleran, busulfan), the epoxides such as 1,2,3,4-diepoxybutane 8 and ethyleneimines (aziridines) such as 2,4,6-tris(1-aziridinyl)-S-triazine (Tretamine, T.E.M. 9).

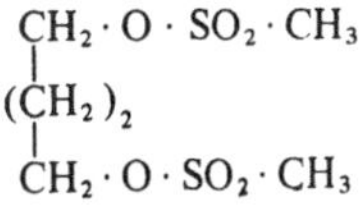

$CH_2 \cdot O \cdot SO_2 \cdot CH_3$
$(CH_2)_2$
$CH_2 \cdot O \cdot SO_2 \cdot CH_3$

7

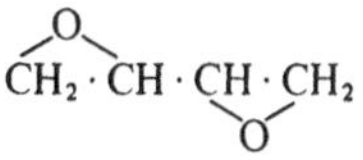

$CH_2 \cdot CH \cdot CH \cdot CH_2$ (with O epoxide bridges)

8

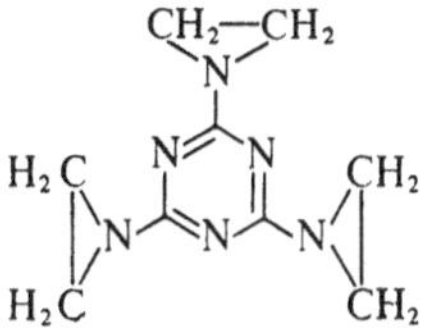

9

More recently a group of halogenated sugars have been found to have similar properties to the bifunctional alkylating agents[11]. 1,6-Dibromo-1,6-dideoxy-D-mannitol, for example (DBD, Mitobronitol; 10) is now in clinical use, but it is believed to act at least in part by transformation in vivo to the bis epoxide 11[12].

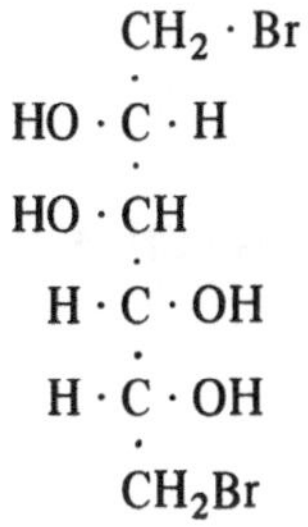

$CH_2 \cdot Br$
$HO \cdot C \cdot H$
$HO \cdot CH$
$H \cdot C \cdot OH$
$H \cdot C \cdot OH$
CH_2Br

10

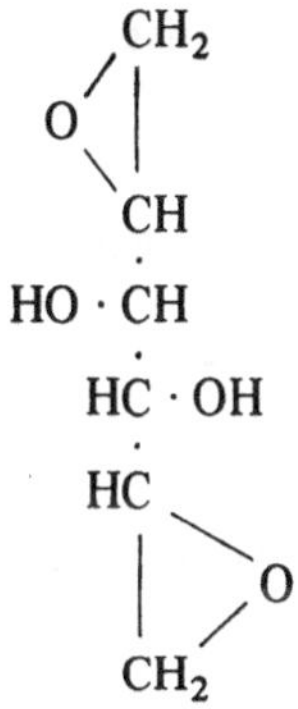

CH_2 (O epoxide)
CH
$HO \cdot CH$
$HC \cdot OH$
HC (O epoxide)
CH_2

11

Later on other types of anti-cancer agent were discovered which showed some resemblance to the above alkylating agents and which could also form electrophilic reactants in vivo. However, they are sufficiently different in many of their chemical and biological properties for them to be put in a separate category from the classical difunctional alkylating agents. The most important of these are the anti-tumour nitrosoureas such as N,N'-bis-2-chloroethyl-N-nitrosourea (BCNU; 12), the triazenes such as 5-(3,3-dimethyl-1-triazeno)imidazole-4-carboxamide (DIC; 13) and hydrazine derivatives such as N-isopropyl-α-(2-methylhydrazino)-p-toluamide (Natulan, procarbazine; 14). The cis-dichloro-diamine platinum compounds, for example cis-dichlorodiammine platinum(II) 15, also show a close similarity to the alkylating agents in their biological properties and their mechanism of action is probably analogous to alkylation.

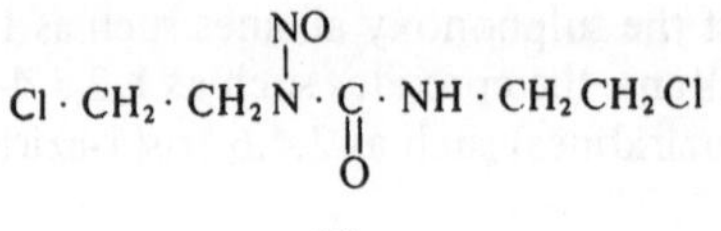

$$\underset{12}{\text{Cl} \cdot \text{CH}_2 \cdot \text{CH}_2 \overset{\text{NO}}{\underset{\|}{\text{N}}} \cdot \underset{\text{O}}{\overset{}{\text{C}}} \cdot \text{NH} \cdot \text{CH}_2 \text{CH}_2 \text{Cl}}$$

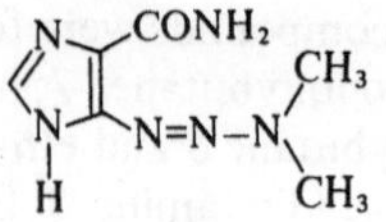

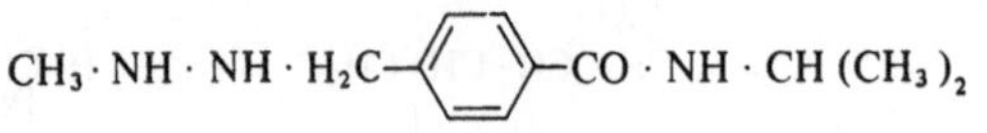

13

$$\underset{14}{\text{CH}_3 \cdot \text{NH} \cdot \text{NH} \cdot \text{H}_2\text{C} - \!\!\!\left\langle \!\!\!\bigcirc\!\!\! \right\rangle\!\!\! - \text{CO} \cdot \text{NH} \cdot \text{CH} (\text{CH}_3)_2}$$

Mechanisms of Alkylation

The chemistry of alkylation has been studied in particular detail for the difunctional alkylating agents with anti-tumour activity. In most cases alkylation proceeds through a second order nucleophilic substitution (S_N2) as exemplified by the ethyleneimines:

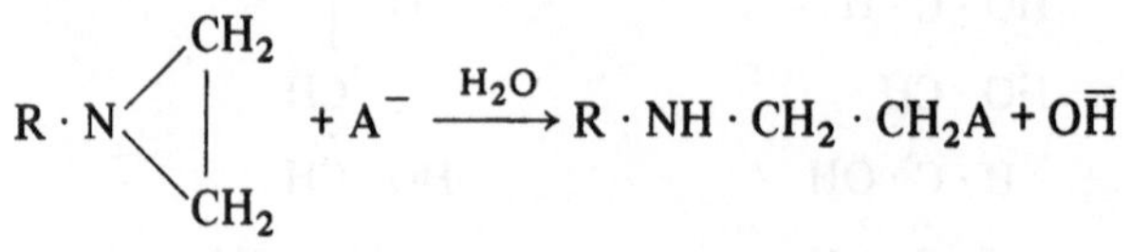

$$\text{R} \cdot \text{N} \underset{\text{CH}_2}{\overset{\text{CH}_2}{\diagup\!\!\!\!\mid\!\!\!\!\diagdown}} + \text{A}^- \xrightarrow{\text{H}_2\text{O}} \text{R} \cdot \text{NH} \cdot \text{CH}_2 \cdot \text{CH}_2\text{A} + \text{O}\bar{\text{H}}$$

or the epoxides;

$$\text{R} \cdot \underset{\diagdown_{\text{O}}\diagup}{\text{CH} - \text{CH}_2} + \text{A}^- \xrightarrow{\text{H}_2\text{O}} \text{R} \cdot \underset{\text{OH}}{\text{CH}} \cdot \text{CH}_2\text{A} + \text{O}\bar{\text{H}}$$

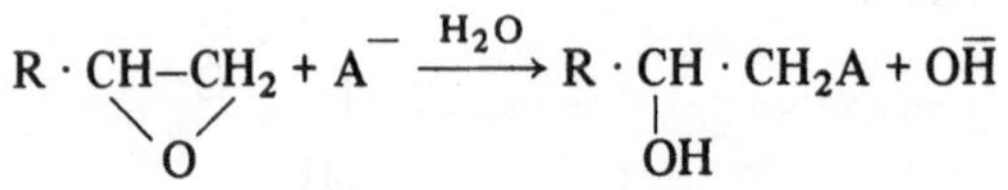

Since both classes of agent act by a bimolecular mechanism their rate of reaction with nucleophilic centres will be dependent on the concentration of such centres. Both epoxides and ethyleneimines are more reactive under very acid conditions. Ethyleneimides such as triethylenephosphoramide (TEPA; *16*) are also more reactive than ethyleneimines because of the electron withdrawing properties of the oxygen atom which makes the methylene group of the ethyleneimine ring more susceptible to nucleophilic attack.

15
$$\underset{\text{H}_3\text{N}}{\overset{\text{H}_3\text{N}}{\diagdown}}\text{Pt}\underset{\text{Cl}}{\overset{\text{Cl}}{\diagup}}$$

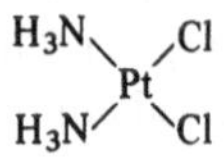

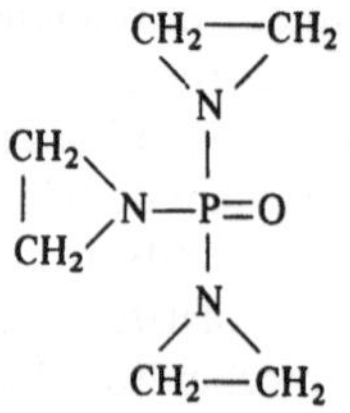

16

The aliphatic nitrogen mustards act similarly after formation of a cyclic immonium ion. The unimolecular conversion to the immonium ion is relatively fast and once formed it reacts by an S_N2 mechanism at a rate dependent on the concentration of nucleophilic centres;

$$R_2N \cdot CH_2 \cdot CH_2Cl \rightleftharpoons R_2\overset{+}{N} \Big\langle \begin{array}{c} CH_2 \\ | \\ CH_2 \end{array} + \bar{Cl} \xrightarrow{\bar{A}} R_2N \cdot CH_2 \cdot CH_2 \cdot A.$$

There is some question as to the mechanism of alkylation of the less basic aromatic nitrogen mustards and sulphur mustard. In these compounds the lower basicity of the nitrogen or sulphur atom does not allow the formation of a stable cyclic immonium ion analogous to the aliphatic nitrogen mustards. They also show first order kinetics in that the rate of alkylation is often independent of the concentration of nucleophilic centres. This can make important differences in their biological properties. The toxicity of aliphatic nitrogen mustards for instance, can be reduced by prior administration of thiosulphate, which increases the rate of reaction of the alkylating agent before it reaches sensitive tissues. There is no similar increase in the rate of reaction of aromatic nitrogen mustards in extracellular fluid, and their toxicity cannot be reduced by thiosulphate.[13]

The finding that for many aromatic nitrogen mustards chloride and hydrogen ions were always simultaneously released on hydrolysis, led to the suggestion that the aromatic nitrogen mustards reacted through a carbonium ion formed by unimolecular loss of chloride ion rather than by formation of a cyclic intermediate analogous to the aliphatic nitrogen mustards[14, 15].

$$R_2N \cdot CH_2 \cdot CH_2^+$$

Slow   Fast

$$R_2N \cdot CH_2 \cdot CH_2Cl \qquad\qquad R_2N \cdot CH_2 \cdot CH_2A$$

Fast   Slow

$$R_2\overset{+}{N} \Big\langle \begin{array}{c} CH_2 \\ | \\ CH_2 \end{array}$$

It has never been entirely accepted that the aromatic nitrogen mustards act exclusively through a carbonium ion intermediate[16, 17] and for some aromatic nitrogen mustards, chloride may be released at a significantly faster rate than alkylating activity disappears, implying the formation of a relatively stable, reactive intermediate[18]. However, there is no doubt that for many aromatic nitrogen mustards there is no accumulation of a cyclic intermediate and the rate determining step is first order ionisation of the halogen atom.

In the sulphonoxyalkane series, some members alkylate predominantly by an S_N1 mechanism and others by an S_N2 mechanism[19].

Biological Properties of Alkylating Agents

Animals dying from the toxic effects of alkylating agents usually show bone marrow hypocellularity, generalised lymphoid depletion and damage to intestinal mucosa and other epithelial surfaces[20, 21]. Death may be a result of the loss of body fluids due to intestinal mucosa damage or of the sequelae of bone marrow damage. Other side effects that may occur include alopecia, bladder cystitis, kidney damage and effects on the central nervous system. Individual agents may have characteristic side effects. For instance, N,N-di-2-chloroethyl-N'-O-trimethylenephosphordiamide (Cyclophosphamide, Endoxan; *17*) causes severe alopecia when administered for the first time and cystitis, presumably because it has a relatively sparing action on the bone marrow and platelets which allows the administration of high doses. Young rats injected with certain nitrogen mustards containing carboxylic acid substituents often die, within minutes, with convulsions. Survivors later show first damage to the intestinal mucosa (four to six days after injection) and later bone marrow damage (usually grossly apparent about 5−10 days after injection). For some alkylating agents such as myleran *7* the bone marrow is the most sensitive site and at minimum lethal doses deaths in rats will occur around 11−15 days. Other alkylating agents have a greater effect on the intestinal mucosa of rats and the majority of deaths will be between four and six days, except for those which kill acutely, within minutes. Although there may be variations in particular side effects, the alkylating agents are broadly similar in their toxicology and there seems no doubt that all act by a similar mechanism of action. The only possible exceptions are myleran and its 1,4-dimethyl derivative (dimethyl myleran; *18*) which are sufficiently different in many of their biological properties for them to be classified separately[22].

$$Cl \cdot CH_2 \cdot CH_2 \diagdown N-P \diagup O-CH_2 \diagup CH_2$$
$$Cl \cdot CH_2 \cdot CH_2 \diagup \quad NH-CH_2$$

17

$$CH_3 \cdot SO_2 \cdot O \cdot CH(CH_2)_2 \cdot CH \cdot O \cdot SO_2 \cdot CH_3$$

18

As with X-irradiation, which the alkylating agents resemble, all types of cell would be damaged at high enough dose levels. Whole animal toxicity, however, is a result of damage to the most sensitive cells, and there is no doubt that with few exceptions, these cells are those that are rapidly dividing. Alkylation is apparently a much more toxic event if cells are forced to divide a short period after administration of an alkylating agent. Thus growing root tips and shoots of plants and germinating seedlings are the most sensitive vegetable cells[20, 23], while bacteria in exponential growth are more affected than when growing slowly in minimal media[24]. Similar observations have been made in a variety of systems ranging from bacteria, yeasts, higher plants and invertebrates to vertebrates including primates. In all cases, at physiological dose levels, tissues with a high mitotic index are the most sensitive. Cell death is apparently the results of unbalanced growth, RNA and protein synthesis being out of phase with DNA synthesis. Other agents can act in a similar way[25, 26] and the process is similar to 'thymine-less' death described in bacteria[27].

In recent years it has been shown that dividing cells pass through a number of distinct phases between each division. Immediately after mitosis, each daughter cell undergoes a period of protoplasmic growth during which there is RNA and protein synthesis and a build-up of the various products required for normal cell function. This period, the G_1 is then followed by a DNA synthetic period (S) devoted almost entirely to DNA replication. The various products necessary for this period of DNA synthesis have been prepared at the latter part of the G_1 period, sometimes referred to as the late G_1. After duplication of the DNA content of the cell, there is a relatively short period, the G_2, during which the cell prepares for the final mitotic phase of the cell cycle (M). It has been found that some anti-cancer agents are phase specific, killing only those cells which are in a particular phase of the cycle. The majority of other anti-cancer agents, including the alkylating agents, are proliferation dependent, again acting only on cells in cycle but at all phases of the cycle[28, 29]. Many tissues comprise cells which remain in G_1 and thus are not sensitive to anti-cancer agents. Such 'resting' cells may also be found in malignant tissue and limit the success of chemotherapy[30]. After treatment has removed all dividing cells the resting cells which escape treatment may then enter the cell cycle and proliferate to reform the tumour.

Mechanism of Action of Alkylating Agents

Electrophilic reactants under the appropriate conditions react with any molecule containing a nucleophilic centre such as an ionised hydroxyl, thiol, phosphate or carboxylic acid group or an uncharged amine. The observation that the alkylating agents react co-valently with a variety of cell molecules even after administration of small doses is not therefore surprising. The identification of such molecules is of interest but does not necessarily prove how these compounds exert their cytotoxicity. Thus early studies on the alkylating agents showed that many enzymes reacted co-valently, but at physiological dose levels the total alkylation was small compared with the number of enzyme molecules present, so that a serious enzyme inhibition did not necessarily result[15]. At present the most satisfactory working hypothesis is that DNA is most sensitive to alkylation and it is some aspect of inhibition of DNA function that is the primary cause of cytotoxicity. Because of its high molecular weight, the alkylation of DNA expressed on a molar basis is far higher than the alkylation of protein, RNA or any other macromolecule. The assumption of DNA as the target site neatly explains why bifunctional agents are so much more cytotoxic for tumours than monofunctional analogues. Only 'the former can cross-link DNA and it is proposed that this is the principle cause of mitotic inhibition[31, 32].

There are several regions of the DNA molecule likely to be alkylated and although at one time it was thought that cross-linking of guanine moieties of adjacent DNA strands in the N_7 position was the predominantly toxic reaction, this view has now been modified and it is accepted that inter or intrastrand cross linking of a number of nucleophilic centres occurs and all may well contribute to cytotoxicity. Following alkylation of a purine base in DNA, the base may be lost from the chain which is then less stable and may break down further. In cross linked DNA, the breakdown is likely to be more serious since after depurination in adjacent regions both strands would be unstable and main chain scission would be likely to occur. This difference between monofunctional and difunctional alkylating agents can adequately explain the different biological properties of the two classes and is the basis of the evidence for DNA as the essential target for the latter compounds. Monofunctional alkylation is less destructive but can cause sub-lethal damage to DNA, leading to mutations, including carcinogenic changes. It is well established that, although the monofunctional agents are not good tumour inhibitors, they are at least as efficient as difunctional compounds in their mutagenic and carcinogenic action.

The sensitivity of DNA to alkylation receives support from a wide variety of experiments which have shown, among other things, that DNA viruses are more susceptible than RNA viruses and that DNA synthesis in many cell lines is inhibited at dose levels that have no effect on RNA or protein synthesis. There is also a correlation between the sensitivity of mammalian and bacterial cells growing *in vitro* to alkylating agents and their ability to repair alkylated DNA. Other mechanisms of action have been proposed to account for the effects of individual alkylating agents[33-35]. While these may play a role, it is quite clear that cells with acquired resistance to one alkylating agent are cross resistant to all others and indicates that all act by a common pathway.

Difunctional Alkylating Agents with Anti-Tumour Action

The nitrogen mustards were the first chemicals to have significant anti-tumour activity in man, and their discovery was made at a time when there was an expanding effort in cancer research. Many of the problems of cancer chemotherapy such as the poor predictability of animal models and the variable response of the different types of human tumour to treatment were not realised. It was the optimistic aim to discover a chemical with highly selective action against all classes of cancer. Not surprisingly, a considerable number of alkylating agents were synthesised, almost at random, and tested for their anti-cancer activity in animal models[37-39]. It was, however, soon realised that the majority of alkylating agents had very similar anti-tumour properties and rational approaches were soon developed in attempts to select more useful agents. It was found, for instance, that there was a correlation between chemical reactivity and anti-tumour action[15, 19]. If agents were very chemically reactive they often had no anti-tumour effects when administered systemically, because they reacted completely before reaching the tumour. On the other hand, a chemical with very low reactivity was excreted before a sufficiently high level of alkylation could take place and, although non-toxic, was also not an effective anti-tumour agent. An optimum level of chemical reactivity therefore existed for a compound administered systemically. Sulphur mustard, for example, with a half-life for hydrolysis of only a few minutes, does not effect the Walker carcinoma *in vivo*, nor do non-metabolisable azo-mustards with a half-life of many hours. The majority of the clinically useful nitrogen mustards, such as aniline mustard (4), melphalan (p-(di-2-chloroethylamino)-L-phenylalanine; *19* and chlorambucil (p-(di-2-chloroethylamino)phenylbutyric acid; *20* have half-lives for hydrolysis of between 30 and 90 min.

$$Cl \cdot CH_2 \cdot H_2C$$
$$Cl \cdot CH_2 \cdot H_2C$$ N— —CH$_2$·CH \quad NH$_2$ / COOH

$$Cl \cdot CH_2 \cdot H_2C$$
$$Cl \cdot CH_2 \cdot H_2C$$ N— —CH$_2$·CH$_2$·CH$_2$·COOH

19 \qquad\qquad\qquad *20*

This correlation between chemical reactivity and anti-tumour action later enabled a rational approach in the design of anti-tumour agents. On the one hand, mustards were synthesised with half-lives of only seconds, for administration directly into arteries supplying the tumour and, on the other hand, agents were prepared with very long half-lives and therefore inactive but which might be selectively metabolised in the tumour to active compounds.

Clinically Useful Alkylating Agents

Despite the synthesis and testing of many thousands of alkylating agents, only a handful are in general clinical use, and HN2 *3*, one of the first to have a clinical trial, is still used, particularly in the treatment of lymphomas, especially Hodgkin's disease[40, 41]. Melphalan *19* has been widely used in the treatment of melanoma and from time to time against a number of other tumours which also respond to cyclophosphamide, especially myeloma. Chlorambucil *20* is a frequent treatment for chronic lymphocytic leukaemia and is also used against ovarian carcinoma. Cyclophosphamide *17* is the most popular nitrogen mustard in cancer chemotherapy and is claimed to have a wider spectrum of action than other alkylating agents, being the only one, for instance, to be used in combination with other agents in the treatment of acute lymphoblastic leukaemia. A number of epoxides have had clinical trials over the years but none have displaced the nitrogen mustards mentioned above. However, the finding that many mannitol derivatives may form dianhydro structures *in vivo*, has led to a renewed interest in epoxides and some diepoxides derived from mannitol have had clinical trials recently. Triethylene melamine (TEM; *9*) had high selectivity against many animal tumours but became discredited as a clinically useful agent because of its variable toxicity. Like most ethyleneimines, it is more reactive under acid conditions and therefore subject to acid catalysed breakdown in the stomach, depending on the amount of acid present. Although this variability could be partially avoided by simultaneous administration of sodium bicarbonate to fasting patients, it is no longer used. ThioTEPA (sulphur analogue of *16*) has been used in the topical treatment of surface tumours but has no special advantage over the more generally accepted agents.

The sulphonoxyalkane myleran (busulfan; *7*) has distinct biological properties causing a particularly pronounced suppression of granulocytes. Mainly for this reason it has been traditional choice for chronic myeloid leukaemia but more recently other alkylating agents have proved to be just as effective in the treatment of this condition.

The earliest alkylating agents were assessed as single agents in clinical trials and if the trial was adequate it was possible to determine whether a new agent had advantages over one in clinical use. However, for some tumours more than one class of chemotherapeutic agent may be effective, and it has been found that with certain drug combinations the results of treatment are far superior to those obtainable using single agent chemotherapy. A new alkylating agent can only be tested against such tumours as part of a drug combination, replacing the conventional alkylating agent, and the comparison between the two agents is more difficult to make. New alkylating agents will continue to be synthesised but only if there is a good reason that they will be more selective clinically than presently available analogues.

Selectivity of Action of Alkylating Agents

The alkylating agents do not kill cancer cells exclusively but under certain conditions they can cause complete regression of established tumours in animals and man at tolerated dose levels[42, 43]. A variety of factors may operate to ensure that an injected alkylating agent reacts with essential molecules in the cancer cell to a greater extent than in sensitive host cells. Drug uptake is an important factor and a cytotoxic agent with an affinity for tumour tissue would be more selective than one taken up equally by all cells. Metabolic pathways are known which can detoxify some alkylating agents and if only normal cells had such pathways they would obviously be less sensitive to alkylation than malignant cells. Alternatively, metabolism might chemically activate some alkylating agents and cause them to be more toxic and, in this case, it would be an advantage for the metabolic pathway to be a unique feature of the tumour cell. Alkylating agents may be prevented from reaching DNA by reaction with unimportant molecules and the concentration of these molecules may vary from cell to cell playing a role in determining the sensitivity of the cell to alkylation. All cells also have mechanisms for repairing DNA damaged by alkylation[44], and the loss of this process by tumour cells would make them highly susceptible to alkylating agents.

Any of these factors can ensure that at tolerated dose levels there is a high tumour cell kill and perhaps complete tumour eradication.

Selective Uptake by Tumour Cells

The distribution, metabolism and excretion of many alkylating agents have been studied in mammals and less frequently in man[21, 45]. Generally they show no selective localisation in sensitive organs such as tumour and bone marrow. There is also a uniform distribution amongst cellular particles and macromolecules with alkylation of many different sites. On injection, alkylating agents distribute fairly uniformly throughout the body with the highest concentration usually occurring in kidney and liver. The tumour, particularly if it is a large mass with a poor blood supply to inner areas, usually takes up less drug despite the fact that it may be sensitive to treatment. Fig. 1 shows the uptake of various chemicals by the Walker carcinoma and liver.

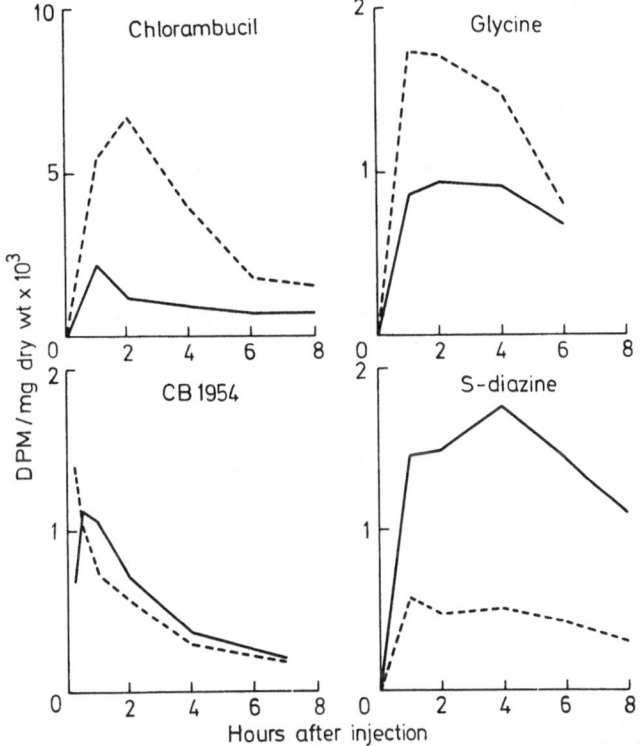

Fig. 1. Uptake of small molecular weight material by the solid Walker carcinoma (_____) in rats compared with liver (_ _ _ _ _). Chlorambucil and CB 1954 cause complete tumour regression but neither are taken up selectively by the tumour. In contrast sulphadiazine shows a much greater concentration in the tumour than in the liver

Very few compounds attain a high concentration throughout the tumour, even though some like chlorambucil can cause complete tumour regression.

151

An exception is sulphadiazine which concentrates selectively in the tumour[46]. This is probably a result of the low pH of the tumour caused by its high anaerobic glycolytic rate and production of lactic acid[47]. At this lower pH more sulphadiazine is precipitated as the unionised form, which is very insoluble, than in cells at physiological pH. Sulphadiazine is not an anti-tumour agent in its own right, but it has been used as a carrier to transport cytotoxic nitrogen mustard groups (sulphadiazine mustard; 21) and triazenes (sulphadiazine triazene; 22). Although the former compound had the requisite physico-chemical properties, if did not concentrate in tumours in the same way as sulphadiazine. The latter, however, does concentrate and appears, in preliminary experiments, to be a highly selective anti-cancer agent.

Cl·CH$_2$·H$_2$C, Cl·CH$_2$·H$_2$C >N—⟨ ⟩—SO$_2$NH—⟨N pyrimidine⟩

21

H$_3$C, H$_3$C >N—N=N—⟨ ⟩—SO$_2$·NH—⟨N pyrimidine⟩

22

Other attempts to exploit the pH difference between malignant and normal cells have involved the design of agents with basic aliphatic side chains which should concentrate in the tumour cell and the use of agents chemically activated at low pH[48, 49].

Concentration in tumours may also be achieved by designing alkylating agents such as 23 which have a half life of only a few seconds and injecting them directly into the afferent vessels supplying the tumour[50].

$$\text{HOOC—⟨ ⟩—O·CH}_2\text{·CH·S·CH}_2\text{·CH}_2\text{Br}$$
$$\text{CH}_2\text{·S·CH}_2\text{·CH}_2\text{Br}$$

23

Such compounds react within the tumour bed but hydrolyse before reaching sensitive areas of the body such as bone marrow. 23 has been used clinically in the treatment of head and neck cancer[51]. A related approach is to design alkylating agents which are chemically unreactive, but which can be activated in the appropriate tissue to form derivatives of such a short half-life that they react only in the tissues where they are activated. This is the principle behind the design of compounds for the treatment of primary liver cancer[52]. The administered alkylating agent is an azomustard 24 which is non-toxic and not chemically reactive. Malignant hepatocytes have a high level of azo-reductase and reduce the azo-compound to the amine 25 which alkylates readily but only has a half-life of 41 sec. Alkylation will therefore occur predominantly in the liver, including the malignant liver cells, and the bone marrow and other sensitive normal tissues will be protected.

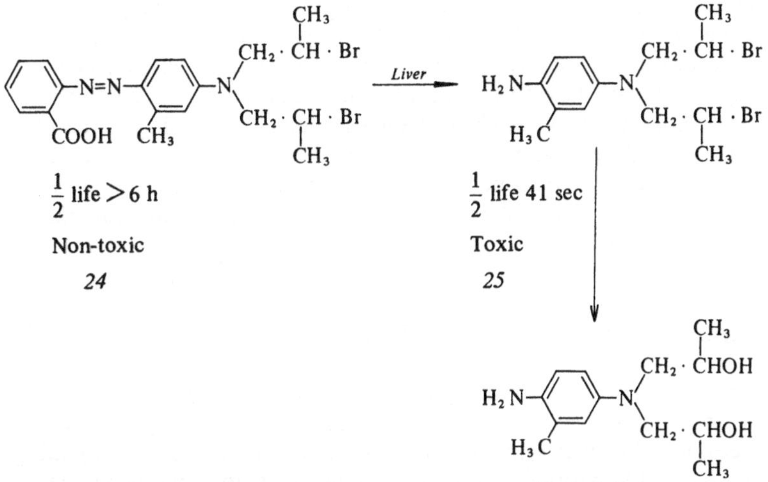

$\frac{1}{2}$ life > 6 h

Non-toxic

24

$\frac{1}{2}$ life 41 sec

Toxic

25

Non-toxic
hydrolysis
product

Selective Activation by Tumour Cells

The principle behind the design of latently active alkylating agents was that they should be activated in tumour cells selectively. Compared with most other cells, however, tumours tend to be low in drug metabolising enzymes and compared with their tissue of origin have lost rather than gained particular enzymes. The chances of discovering alkylating agents selectively activated by tumours are therefore not good, but in rare instances this pathway can be responsible for the good anti-tumour effects obtained with an alkylating agent. Advanced plasma cell tumours in mice undergo complete regression after small doses of aniline mustard (4) but not related agents. It was eventually shown[42, 53, 54] that on injection, aniline mustard was converted to a large extent to the O-glucuronide *26* in the liver. This was a detoxification reaction since the glucuronide, later prepared synthetically, had an LD_{50} some three times higher than that of the parent compound.

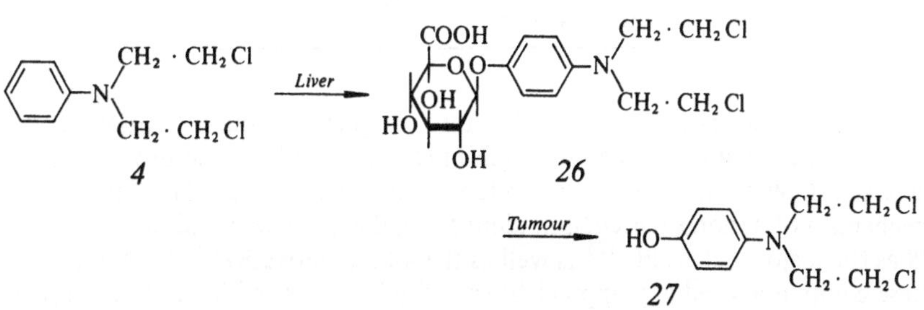

However, sensitive tumours had abnormally high levels of β-glucuronidase, some of it extra lysozomal, which converted the glucuronide to the highly toxic p-hydroxyaniline mustard 27 causing selective tumour cell kill. The clinical application of these findings consist of administering aniline mustard to patients whose tumours have been shown at biopsy to have high levels of glucuronidase[55]. Extension of this work would be the use of the O-glucuronide of aniline mustard in the treatment of cancers shown to be high in β-glucuronidase and the use of related phosphates and sulphates against tumours high in sulphatase and phosphatase.

Selectivity by Deactivation in Normal Tissues

Since tumours, particularly rapidly growing ones, tend to lose enzymes, it is likely that selectivity may be achieved by the use of an alkylating agent that can be detoxified by an enzyme only present in normal cells. Some alkylating agents can be readily detoxified *in vivo* by known metabolic pathways. HN2, for instance, is demethylated in liver to the less active nor-HN2[56], and aromatic nitrogen mustards can be converted to their less toxic monofunctional analogues[56, 57]. There is no evidence, however, that any of these mechanisms do protect bone marrow cells from the alkylating agents since the enzymes responsible occur mainly in the liver.

It has recently been proposed that the high selectivity of cyclophosphamide may be due to selective detoxification of a potentially cytotoxic metabolite in normal tissue[58]. Originally prepared as one of a series of latent alkylating agents that might be activated by tumours[59], it was soon found that cyclophosphamide was activated predominantly by an initial metabolic step in the liver[60]. Experiments of the kind shown in Table 1 demonstrated that while

Table 1. Activation of cyclophosphamide by liver microsomes

	Dose to kill 90% tumour cells
Walker tumour *in vivo*	20 μg/ml
Walker tumour *in vitro*	6,000 μg/ml
Walker tumour *in vitro* + liver microsomes + NADPH	10 μg/ml

low concentrations of the drug could cause complete tumour inhibition *in vivo*, it was innocuous to the same cells *in vitro*, unless liver microsomes and an NADPH generating system were added. Extensive work to elucidate the mechanism of action succeeded in identifying the propionic acid derivative 28 as the major metabolite[62] as well as the 4-keto derivative[63] 29. None of these compounds had any appreciable activity however and it was later shown that acrolein *30* and phosphoramide mustard *31* were produced on incubation of cyclophosphamide with liver microsomes[64, 65]. These facts plus the identification of a derivative of 4-hydroxycyclophosphamide *32*[58)] led to the following scheme being postulated for the selective action of cyclophosphamide.

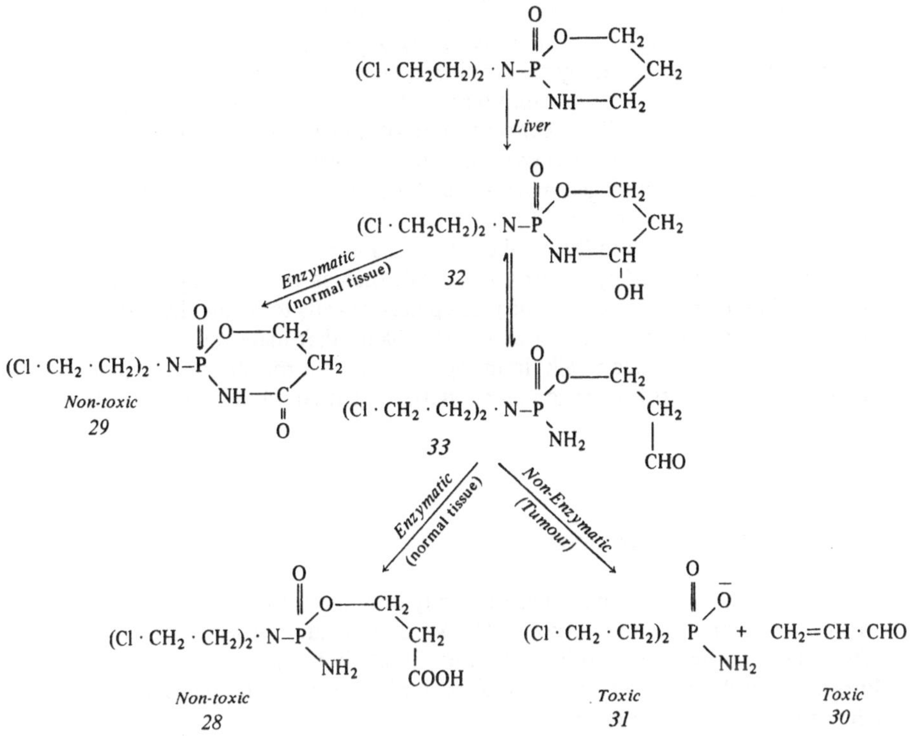

The initial metabolism of cyclophosphamide is in the liver to the 4-hydroxy-derivative which is in equilibrium with the tautomeric aldehyde *33*. It is proposed that selective anti-tumour action would result if normal cells could detoxify this metabolite by enzymatic conversion to known excretory products, the 4-keto or propionic acid derivatives *24* and *28*. In the absence of such enzymes the metabolite is known to be unstable and to break down to acrolein *30* and phosphoramide mustard *31*, both of which are very toxic to cells in culture[66]. Selective formation of these toxic breakdown products in tumours that have lost the appropriate detoxifying enzymes could obviously result in toxicity.

The proposed scheme requires confirmation, 'however, the 4-hydroxy metabolite has recently been synthesised and should enable the theory to be tested[67].

Selectivity from Competing Nucleophiles and Defective Repair

If DNA is the target site for alkylating agents then cytotoxicity will presumably be related to the level of permanent DNA alkylation. Even though two cells take up equal quantities of an alkylating agent, the level of DNA alkylation may differ considerably if one cell but not the other has a high concentration of a non-essential nucleophile which competes with the DNA for alkylation. It has been shown many times that, by increasing the thiol content of a tissue, the sensitivity to alkylating agents is reduced because of a smaller amount of DNA alkylation[68, 69]. The inherent sensitivity of a tumour may therefore be dependent on the intracellular concentration of non-essential nucleophiles. In this connection it has been shown that the response of a range of transplanted tumours to a nitrogen mustard can be correlated with the ratio of free to protein-bound thiol content[70].

The repair of damaged DNA takes place as a normal cell function and in its absence, pathological symptoms may result such as xeroderma pigmentosa, which is characterised by an inability of epidermal cells to repair thymidine cross-linking caused by U.V. irradiation. It is likely that malignant tissues which are highly sensitive to alkylating agents have lost the ability to repair DNA efficiently. In this situation an alkylating agent could be highly selective.

Development of Resistance

The bifunctional anti-tumour agents actually only inhibit the growth of a small proportion of animal tumours. Many, some rapidly growing, are not effected even at maximum-tolerated doses. Similarly in man, although Burkitt's lymphoma is very sensitive to cyclophosphamide and merophan (the ortho-racemic analogue of melphalan), the more common solid tumours, the carcinomas of the lung and digestive tract, do not respond at all the alkylating agents. Even tumours which are initially sensitive to alkylating agents may become resistant on treatment. The study of acquired resistance is of obvious importance since it may lead to ways of overcoming the development of resistance and thus allow continuous treatment until every malignant cell has been eradicated.

Resistance can theoretically arise by a number of mechanisms[36, 71] and generally they are the reverse of the conditions for selectivity just described. Resistance could thus be a failure to concentrate a drug or to activate a latent agent. Conversely, it could be the outgrowth of cells with a high level of non-essential nucleophiles or of the appropriate drug detoxifying enzyme.

However, where quantitative studies have been made on tumours with acquired resistance to a particular alkylating agent, it is a general finding that such tumours are then cross resistant to all other difunctional agents[36, 72]. There must therefore be some basic mechanism of resistance which can work equally well for all agents of this class. Because of the variety of structures involved, it is difficult to envisage a mechanism whereby the enzymatic deactivation of all agents could occur. Monodechloroethylation can convert difunctional nitrogen mustards to the inactive monofunctional derivatives, but it is unlikely that such an enzyme could also act on ethyleneimines and epoxides. A possible common mechanism could be an increase of reactive nucleophiles close to the sensitive sites of DNA, thus protecting it from alkylation. This could be achieved perhaps by the thiol groups of a nucleoprotein acting as the protecting nucleophile, but there is no good experimental evidence for this. An induction of a DNA repair mechanism would represent a general pathway for the acquisition of resistance and such a pathway has been shown to be present in mammalian cells[73]. DNA can be repaired in a number of ways[44] but for alkylated DNA the so-called excision repair is most likely. This is initiated by an endonuclease which makes a cut in the DNA strand close to the area of alkylation. A portion of the strand, including the damaged region is then removed by an exonuclease. Subsequently a replacement section is synthesised by a polymerase and fitted into the strand by a ligase enzyme.

Resistance can be due to selection of cells already present in the tumour and which survive alkylation, or a result of enzyme induction in non-lethally alkylated cells. If excision repair is the major cause of resistance, then sensitivity might be restored by interfering with the repair process, and some progress has already been made in this direction[74].

Triazenes with Anti-Tumour Activity

3,3-Dimethyltriazenes were first shown to have anti-tumour activity a number of years ago[75], but the development of a clinically useful agent came from attempts to design antagonists of 4-aminoimidazole-5-carboxamide (AIC) whose ribotide is a precursor in purine biosynthesis. 5(3,3-Dimethyl-1-triazene)imidazole-4-carboxamide (DIC; 13) is probably the most effective agent in the treatment of melanoma[76]. There is no evidence however that DIC is an antagonist of purine biosynthesis and other triazenes not containing the imidazole ring are just as effective as DIC in inhibiting tumour growth[77], the essential requirement being the presence of a triazeno group containing at least one methyl substituent in the 3 position (Table 2). Although the triazenes are at the moment only of limited clinical use, they have shown a remarkable specificity of action against a variety of animal tumours. From their spectrum of action it is clear that they differ from the difunctional alkylating agents (Table 3). They are very effective against tumours such as the plasma cell tumour which responds to many alkylating agents but also active against the TLX 5 lymphoma which is completely unresponsive to all difunctional alkylating agents. In their spectrum of action the triazenes closely resemble the nitrosoureas and it is likely that both act by similar pathways.

Table 2. Effect of various triazenes on the growth of the TLX 5 lymphoma.
Survival time of the animal is related to the number of viable tumour cells injected and thus the % increase in survival time (% IST) is a direct measure of tumour cell kill. There is no correlation between dealkylation and anti-tumour activity. However, at least one 3-methyl group on the triazene is necessary for activity

Compound	Optimal dose	% IST	Toxic dose	% Substrate dealkylation
	16	53	128	41.6
	16	79	128	21.4
	50	63	200	49.0
	Inactive	Inactive	200	46
	25	67	200	5.3

Table 3. The effect of different classes of anti-tumour agent on some transplanted tumours. The TLX5 (R) is a line with acquired resistance to a triazene and is cross resistant to BCNU. Triazenes and BCNU are active against tumours which do not respond to anti-metabolites and others that are insensitive to alkylating agents. The platinum complex was cis dichloro-bis(cyclopentylamine)platinum(II) and the triazene 5-(3,3-dimethyl-1-triazeno)-4-carbethoxy-2-methylimidazole

| | Tumour | | | |
	PC6 (TI)	Walker (TI)	TLX5 (% ILS)	TLX5 (R) (% ILS)
Triazene	86	8	67	Inactive
Cyclophosphamide	136	22	Inactive	Inactive
BCNU	96	30	197	19
Platinum complex	200	8	Inactive	Inactive
Methotrexate	Inactive	Inactive	55	84
Cytosine arabinoside	Inactive	Inactive	39	51

Mechanism of Action of Triazenes

Like many other cytotoxic agents the triazenes are also carcinogenic and many of the studies on their mechanism of action have been with the aim of elucidating mechanisms of carcinogenesis[78]. Two metabolic fates of the triazenes are envisaged. In the light or under acid conditions they may breakdown to form diazonium compounds. Since this is predominantly a light catalysed reaction it is unlikely to be of biological significance unless there are regions of tumour cells sufficiently acidic to produce small quantities of diazonium compound.

The diazonium derivatives of certain of the triazenes are stable enough to be isolated and they have been shown to be extremely toxic to many tumours in

159

vitro. However they are not good anti-tumour agents because they are also toxic to whole animals so no selectivity is obtained[77]. An alternative pathway involves demethylation to the mono methyl derivative, which rearranges and subsequently breaks down to methyl carbonium ions. It is postulated that methylation of essential macromolecules accounts for both the anti-tumour and carcinogenic action of these compounds. It is difficult to see how by this mechanism the triazenes kill tumours which are resistant to bifunctional alkylating agents. One would need to assume that methylation is a much more lethal action than the types of alkylation caused by a difunctional agent such as cyclophosphamide. Furthermore, powerful methylating agents such as N-methyl-N-nitrosoguanidine are ineffective against tumours sensitive to triazenes. A further problem is that 3,3-diethyl analogues of the triazenes are just as toxic to cells *in vitro* but have no anti-tumour activity *in vivo*. De-ethylation occurs just as readily as demethylation (Table 2) and presumably the product then breaks down to form diazoethanehydroxide which can alkylate molecules in the same way as the methyl analogue. If methylation is the essential reaction for anti-tumour activity then it must be presumed as a more important reaction than ethylation.

The high anti-tumour selectivity of the triazenes and their broad spectrum of action requires further investigation since it may eventually enable the design of agents for clinical use which maintain their anti-tumour specificity but do not break down into products which are toxic but not tumour inhibitory. The role of formaldehyde produced by the active dimethyl derivatives but not by the inactive diethyl analogues, may be important. It has also recently been proposed that toxic products formed by triazenes may be neutralised by factors in the cytoplasm of normal tissues but not in sensitive tumour cells[77].

Nitrosoureas with Anti-Tumour Activity

The slight activity of 1-methyl-1-nitrosourea (MNU; *34)* against the L1210 leukaemia was discovered as a result of the American National Cancer Institute's large scale screening programme. An investigation of related structures culminated in the discovery of the highly active BCNU *12* and 1-(2-chloroethyl)-3-cyclohexyl-1-nitrosourea (CCNU; *35*) and its *trans*-4-methylderivative (methyl CCNU).

34   *35*

Like the triazenes, the nitrosoureas show many points of resemblance to the difunctional alkylating agents. They inhibit a range of animal tumours sensitive to alkylating agents, but not a hamster tumour with acquired resistance to cyclophosphamide[79]. In their side effects they are similar to the alkylating agents causing bone marrow hypoplasia and a suppression of peripheral blood leucocytes and platelets which is the limiting toxicity in man. Unlike the alkylating agents, the bone marrow damage is not rapid in onset but only occurs in man some 3–4 weeks after administration, lasting about three weeks. BCNU and related nitrosoureas can cross the blood-brain barrier and have been reported effective in the treatment of intracranial metastases

Table 4. Effect of metal complexes on the established plasma cell tumour. The ID_{90} is the minimum dose that causes complete tumour regression. TI = Therapeutic Index

Compound	Activity against the plasma cell tumour		
	LD_{50} (mg/kg)	ID_{90} (mg/kg)	TI
H_3N, Cl / Pt / H_3N, Cl	13.0	1.6	8.1
H_3N, Cl / Pt / Cl, NH_3	27.0	>27.0	<1.0
H_3N, Cl / Pd / H_3N, Cl	50	>50	<1.0
H_3N, Cl / H_3N—Rh—Cl / H_3N, Cl	225	86	2.6
H_3N, Cl / H_3N—Ir—Cl / H_3N, Cl	1500	>1,500	<1.0
H_3N, Cl / H_3N—Ru—Cl / H_3N, Cl	132	>132	<1.0

and in combination with other agents, in the treatment of meningeal leukaemia. BCNU has also been used against Hodgkin's disease and other lymphomas. Methyl CCNU appears to be particularly effective against animal tumours with cell kinetics more closely resembling solid human tumours, possibly because it has an action against cells which are not proliferating. It is therefore at present on clinical trial against a number of solid carcinomas.

Mechanism of Action

Despite obvious similarities to the bifunctional alkylating agents, the nitrosoureas have a completely different spectrum of anti-tumour action. Although little is known about their mechanism of action, they can break down *in vivo*

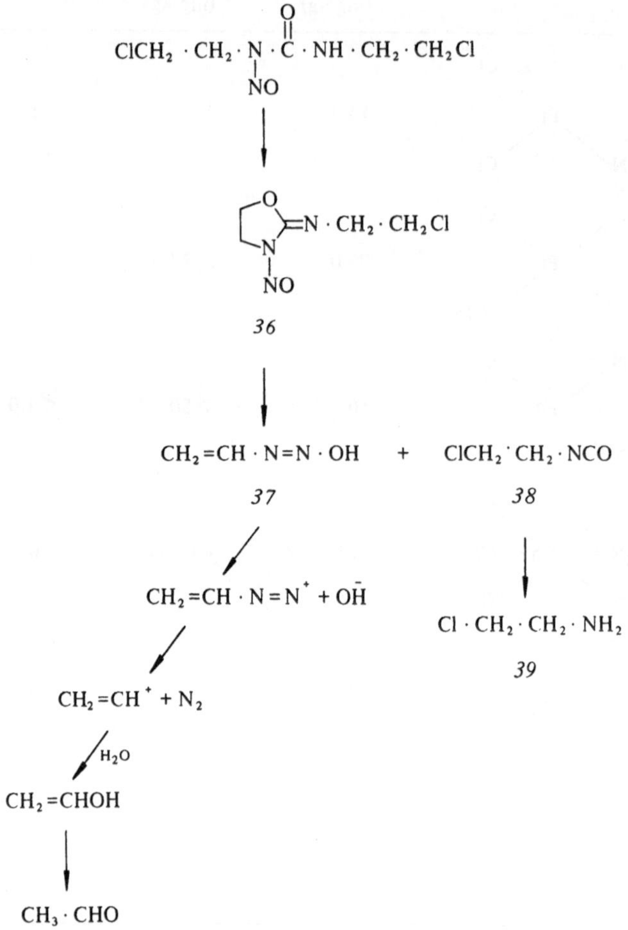

$$ClCH_2 \cdot CH_2 \cdot \underset{\underset{NO}{|}}{N} \cdot \overset{\overset{O}{\|}}{C} \cdot NH \cdot CH_2 \cdot CH_2 Cl$$

36

$$CH_2 = CH \cdot N = N \cdot OH \quad + \quad ClCH_2 \cdot CH_2 \cdot NCO$$

37   38

$$CH_2 = CH \cdot N = \overset{+}{N} + O\bar{H}$$

$$Cl \cdot CH_2 \cdot CH_2 \cdot NH_2$$

39

$$CH_2 = \overset{+}{C}H + N_2$$

$$\xrightarrow{H_2O}$$

$$CH_2 = CHOH$$

$$CH_3 \cdot CHO$$

(unlike the triazenes, no enzymatic activation step is necessary) to a variety of monofunctional alkylating agents. A possible mechanism of action of BCNU involves the loss of HCl to form an oxazolidine *36* which subsequently breaks down to the diazohydroxide *37* and an isocyanate *38*. Acetaldehyde may be formed from the diazohydroxide and the isocyanate may break down to the alkylating 2-chloroethylamine *39*. The latter can also react with unchanged 2-chloroethylisocyanate to form 1,3-bis(2-chloroethyl)urea which is not an active anti-tumour agent. The role of the various breakdown products in determining the anti-tumour action of BCNU is not clear. It is not likely that the acetaldehyde formed is of importance since related fluoroethylnitrosoureas are also active but form fluoroethanol in place of acetaldehyde. The diazo-hydroxide can alkylate as well as the chloroethylamine formed from the isocyanate and these must be considered as important breakdown products. However, no alkylating agent similar to chloroethylamine would be formed from the cyclohexylureas which are just as active as BCNU. The isocyanates which are formed from all active nitrosoureas may play a major role since it has been shown that *in vitro*, the isocyanate from CCNU simulates the parent compound in its cytotoxic properties[79]. Inhibition of protein synthesis by the isocyanates from nitrosoureas is thought to play an important role in tumour inhibition, perhaps combined with monofunctional alkylation. Besides affecting protein synthesis, the nitrosoureas cause a marked inhibition of *de novo* purine biosynthesis and therefore of nucleic acid synthesis. They also interfere with the utilization of histidine in one carbon metabolism through the inhibition of formiminotransferase, and cause alteration in NADase and DNA nucleotidyltransferase activity[80-82]. The high selectivity of the nitrosoureas may be due to a unique combination of properties such as their ability to penetrate certain cells and the breakdown to a number of products acting on different pathways. It seems likely that their mechanism of action is a complex one and may involve attack at more than one site. Alkylation may attack essential molecules like DNA, for example, and carbamoylation of proteins may interfere with the enzymes necessary for its repair.

Platinum Compounds with Anti-Tumour Activity

New classes of anti-tumour agent may, like the nitrosoureas, be discovered by large scale screening, but more often than not they are found by chance, the first observations sometimes being made in the course of experiments not related to cancer chemotherapy. Such was the case with the anti-tumour platinum compounds whose biological properties were first discovered when platinum electrodes were used to apply an alternating electric current across a chamber in which bacteria were growing[83]. The effect of the electric current was to inhibit cell division in the bacteria but not their growth so that normal rod-shaped *E. Coli* formed long filamentous structures. Further research showed that two platinum complexes, formed by electrolysis, were responsible – *cis*-dichlorodiammine platinum(II) and *cis*-tetrachlorodiammine platinum(IV).

163

Subsequently they were found to be powerful inhibitors of the sarcoma 180 in mice[85] and these results have since been confirmed and extended by a large number of workers. Unlike the dimethyltriazenes where the members of the series are very similar in their anti-tumour action provided a monomethyl-triazeno group is present, small changes in structure of the platinum compounds[b] can lead to complete loss of anti-tumour properties. It was established early on (Table 4) that although the *cis* dichloro compound inhibited the growth of animal tumours, the *trans* compound did not. No *trans* dichloro amine has yet had any activity, although being sometimes just as toxic as the active *cis* analogue. Similarly, no other metal can be substituted for platinum, although certain rhodium analogues, some of which are potent anti-bacterial agents, do show some slight activity (Table 4). A series of platinum complexes in which the co-ordination is through atoms other than nitrogen, *e.g.* oxygen and sulphur, have also been examined, but with the exception of *cis*-diammine malanatoplatinum(II) none had activity in tumour bearing animals[86, 87]. The most selective anti-tumour agents based on platinum appear to be cis dichloro diamines, although the nature of the substituent groups on the nitrogen atoms can have a profound effect on anti-tumour activity. The good anti-tumour properties of *cis*-dichlorodiammine platinum are completely lost for instance when a methyl group is placed on the nitrogen atoms *40* even though there is little change in toxicity. Similarly, a β-chloroethyl substituent on each of the nitrogen atoms *41* decreases considerably anti-tumour selectivity. However, a num-

40

41

ber of platinum compounds have now been synthesised which are significantly better against many animal tumours than the parent compound[87]. These chemicals are of two series, one where the ammonia ligands are replaced by compounds containing cyclic nitrogen atoms, *e.g.* *cis*-dichlorobis (ethyleneimine) platinum(II) *42* and another comprising cis dichloro derivatives of platinum co-ordinated with an aromatic or alicyclic diamino group such as dichloro (*o*-phenylenediamine)platinum(II) *43*, or two alicyclic amines.

42

43

---

[b] R. J. P. Williams *et al.*: Structure and bonding *11*, 1 (1972).

The most interesting agents, with a far greater selectivity against a plasma cell tumour than the parent compound, are the alicyclic diamines listed in Table 5. As the alicyclic rings increase in size the toxicity decreases with no change in potency until the dicyclohexylamine derivative, so that the therapeutic index increases with each member of the series. The highest therapeutic index occurs with the cyclohexylamine compound but its anti-tumour potency is low compared with the other members of the series. One of the problems with this class of compound is their poor solubility which makes their formulation for intravenous injection difficult. The most likely candidate for clinical trial will therefore be the cyclopentylamine, since, although it has a lower therapeutic index than the dicyclohexylamine, it is more potent and less will need to be administered.

Table 5

Compound	Activity against the plasma cell tumour		
	LD_{50} (mg/kg)	ID_{90} (mg/kg)	TI
	56.6	2.3	24.6
	90.0	2.9	31.0
	480	2.4	200
	>3,200	12.0	>267
	135	59	2.3

Mechanism of Action of Platinum Compounds

The mechanism of action of the platinum compounds is largely unknown but in many of their biological properties they show a close resemblance to the difunctional alkylating agents. Thus at minimum effective dose levels they cause a selective inhibition of DNA synthesis and also can be shown to cross link DNA [88, 89]. It seems likely that the N_7 and 6 NH_2 positions of adenine are particularly susceptible to attack by these complexes. In DNA there is likely to be cross-linking of complementary strands in areas where there is a sequence of adenine-thymine and thymine-adenine base pairs. In this situation the adenine moieties lie one above another about 3.0 Å apart in an ideal situation for *cis* bidentate cross linking[90]. Many of the techniques previously used to show that DNA was a sensitive target site for alkylating agents have now been applied to the platinum complexes and have confirmed the similarity between alkylating agents and these compounds. Tumours with acquired resistance to an alkylating agent show cross resistance not only to other alkylating agents but also to the platinum complexes[91]. This similarity to alkylating agents is a little disappointing to the cancer chemotherapist, since it implies that any compound used clinically may have the same spectrum of action and similar side effects, and therefore not represent a new class of compound that might be used in combination with present day agents. However, the early clinical results of *cis*-dichlorodiammine platinum(II) do show that it may be useful in the treatment of ovarian cancer with acquired resistance to nitrogen mustards[92]. The limiting toxicity of the platinum complex in man is kidney damage but it is hoped that the more recent analogues will be without this side effect. Of particular interest are the most recently synthesised platinum derivatives with activity against animal tumours. These are the so-called platinum blues which are complexes of various platinum diamine dichlorides with pyrimidines. In contrast to other platinum compounds with anti-tumour properties, these are extremely water-soluble and may well have advantages over the agent at present on clinical trial.

Other Anti-Tumour Agents with Alkylating Properties

Mitomycin C, an antibiotic produced by fermentation of *streptomyces*, has been used extensively in Japan for the treatment of stomach cancer which is prevalent in that country. It probably acts after conversion into an alkylating agent *in vivo*, and it also contains quinone and urethane moieties which may contribute to its anti-tumour effect. A related series of compounds, the pyrollizidine alkaloids, occur in a variety of plants and are known to cause acute liver cytotoxicity when accidentally ingested[93]. Like mitomycin C, these agents are almost certainly metabolised *in vivo* by liver microsomes to alkylating agents which cause the liver toxicity. Some of these alkaloids have anti-tumour properties, presumably because the active metabolite formed in the liver is stable enough to reach the tumour.

$$\overset{+}{N} = N \cdot CH \cdot COO \cdot CH_2 \cdot \underset{\underset{NH_2}{|}}{CH} \cdot COOH$$

44

$$\overset{+}{N} = N \cdot CH \cdot CO \cdot CH_2 \cdot CH_2 \cdot \underset{\underset{NH_2}{|}}{CH} \cdot COOH$$

45

A series of anti-metabolites have also been prepared which contain an alkylating moiety to enable the antagonists to inhibit the appropriate enzyme irreversibly, by formation of a co-valent bond[94]. Two compounds of this type to have had a clinical trial are the glutamine antagonists azaserine *44* and 6-diazo-5-oxo-L-norleucine (DON; *45*).

Both compounds interfere with purine biosynthesis at one of the three stages where glutamine is required. Addition of either of these two inhibitors to cells leads to an accumulation of formylglycineamide ribotide. The inhibition of the enzyme is reversible for a short period but soon becomes irreversible as a result of the inhibitor alkylating the enzyme, probably through a sulphydryl group of a cysteine residue.

N-isopropyl-α(2-methylhydrazino)-p-toluamide (Natulan, Procarbazine; *14*) was one of a series of agents first tested as potential monoamine oxidase inhibitors. The presence of the methylhydrazine residue is essential for anti-tumour activity and it is possible that it requires activation by initial demethylation as has been seen for the dimethyl triazenes. A widely used agent clinically, where it is particularly effective in combination against Hodgkin's disease, surprisingly few studies have been carried out on its mechanism of action[95]. Natulan affects the same normal tissues as alkylating agents, particularly bone marrow. DNA is a possible target site for this compound since it is known to break down under certain conditions to form products which degrade DNA.

Alkylating Agents with Carcinogenic Activity

Although there are a number of theories regarding the cause of cancer, two of the more probable explanations involve a somatic mutation or the activation of a viral genome. In both these cases an agent that alkylates DNA might be the initial stimulus for such a carcinogenic event. Thus alkylating agents which kill cells at high dose levels may be expected to be carcinogenic if administered chronically at lower dose levels when cells are only sub-lethally damaged. The majority of bifunctional alkylating agents that have been administered to rats and mice chronically have, in fact, been shown to be carcinogenic. The mono-functional agents cause less acute cytotoxicity and therefore are sometimes more carcinogenic than their monofunctional analogues because many cells survive alkylation and occasionally transform to a malignant cell. Similarly, the carcinogenicity of triazenes and nitrosoureas has also been well-documented and doubtless the platinum compounds will also be shown to have the same properties when adequately tested.

Alkylation or a related electrophilic reaction, as a prerequisite for carcino-genicity by organic compounds receives support from studies on the mechanism

167

of action of a variety of experimental and environmental carcinogens. 2-Acet-amidoaminofluorene, for instance *46* is a well known hepatocarcinogen and detailed studies of its metabolism *in vivo* have proved that the carcinogenic metabolite is an electrophilic reactant[96]. On injection, it is metabolised by

liver microsomes to a number of products, the majority of which, such as ring hydroxylation, represent detoxification reactions. However, a small proportion is metabolised to N-hydroxy-2-acetamidofluorene *47*. This compound is not the ultimate carcinogen but synthetic esters of it are very powerful electro-philic reactants at neutral pH and the products they form with proteins and RNA are identical with the products isolated from the liver of rats injected with 2-acetamidofluorene. It is likely that once N-hydroxylation has occurred in the microsomes, the sulphate *48* is formed in the cytosol, and it is of interest that animals with high levels of the liver sulphotransferase system are more susceptible to the carcinogenic action of 2-acetamidofluorene than animals low in the enzyme. Reaction with proteins and nucleic acids involves loss of sulphate as SO_4^{2-} to generate the electrophilic reactant *49*. Experimental studies on other chemical carcinogens including azo-benzenes, ethionine,

pyrollizidine alkaloids, nitroquinoline-N-oxides and polycyclic hydrocarbons have identified, in each case, some form of electrophilic reactant as the causative agent. In recent years a number of chemicals have been suspected of causing cancer in man. These include the aflatoxins which originate from the mould, *Aspergillus flavus*, and may contaminate a variety of foodstuffs. It is a likely cause of the very high incidence of hepatocellular carcinoma that occurs in certain areas of Africa where groundnuts, which may be contaminated by *aspergillus*, are a basic constituent of the diet. Other carcinogens have been implicated as contaminants of a variety of plants that may be used as a source of food or which may be accidentally ingested. Amongst these are the glycoside of methylazoxymethanol, a known toxic element of the cycad plants which are used in some countries as a source of edible starch, Safrole used in artificial flavouring and ethionine produced by a variety of enterobacteria. For all these suspected human carcinogens there is growing evidence that, on ingestion, they break down or are converted into electrophilic reactants, similar to the alkylat-ing agents, which react co-valently with cell macromolecules and initiates the carcinogenic transformation.

References

1) Meyer, V.: Ber. dtsch. chem. Ges. *1*, 1725 (1887).
2) Ehrlich, P.: Collected Papers of Paul Ehrlich I (ed. F. Himmelweit), pp. 596–618. London: Pergamon Press 1956.
3) Krumbhaar, E. B., Krumbhaar, H. D.: J. Med. Res. *40*, 497 (1919).
4) Berenblum, I.: J. Path. Bact. *40*, 549 (1935).
5) Adair, F. E., Bagg, H. J.: Ann. Surg. *93*, 190 (1931).
6) Gilman, A.: Am. J. Surgery *105*, 574 (1963).
7) Gilman, A., Philips, F. S.: Science *103*, 409 (1946).
8) Peck, R. M., O'Connell, A. P., Creech, H. J.: J. Med. Chem. *9*, 217 (1966).
9) Hebborn, P., Triggle, D. J.: J. Med. Chem. *8*, 541 (1965).
10) Reist, E. J., Spencer, R. R., Wain, M. E., Junga, I. G., Goodman, L., Baker, B. R.: J. Org. Chem. *26*, 2821 (1961).
11) Sellei, C., Eckhardt, S., Nemeth, L.: Chemotherapy of neoplastic diseases. Budapest: Akademiai Kiado 1970.
12) Elson, L. A., Jarman, M., Ross, W. C. J.: Europ. J. Cancer *4*, 617 (1968).
13) Connors, T. A., Jeney, A., Jones, M.: Biochem. Pharmacol. *13*, 1545 (1964).
14) Ross, W. C. J.: Adv. Cancer Res. *1*, 397 (1953).
15) Ross, W. C. J.: Biological alkylating agents. London: Butterworths 1962.
16) Price, C. C., Gaucher, G. M., Koneru, P., Shibakawa, R., Sowa, J. R., Yamaguchi, M.: Ann. N. Y. Acad. Sci. *163*, 593 (1969).
17) Bardos, T. J., Chmielewicz, Z. F., Hebborn, P.: Ann. N. Y. Acad. Sci. *163*, 1006 (1969).
18) Williamson, C. E., Witten, B.: Cancer Res. *27*, 23 (1967).
19) Stock, J. A. in: Drug design (ed. E. J. Ariens): vol. 2, pp. 531–571. New York and London: Academic Press 1971.
20) Wheeler, G. P.: Cancer Res. *22*, 651 (1962).
21) Ochoa, M., Hirschberg, E.: Alkylating agents in experimental chemotherapy. (eds. R. J. Schnitzer and F. Hawking), pp. 1–32. London and New York: Academic Press 1967.
22) Elson, L. A.: Radiation and radiomimetic chemicals. Comparative physiological effects. London: Butterworths 1963.
23) Skipper, H. E.: The cell cycle and the chemotherapy of cancer, in: The cell cycle and cancer (ed. R. Baserga), pp. 355–387). New York: Marcel Dekker 1971.
24) Papirmeister, B., Davison, C. L.: Biochem. Biophys. Acta *103*, 70 (1965).
25) Rueckert, R. R., Mueller, G. C.: Cancer Res. *20*, 1584 (1960).
26) Jae Ho Kim, Eidinoff, M. L.: Cancer Res. *25*, 698 (1965).
27) Cohen, S. S., Barner, H. D.: Paediatrics *16*, 704 (1955).
28) Van Putten, L. M., Lelieveld, P.: Europ. J. Cancer *7*, 11 (1971).
29) The Cell Cycle and Cancer (ed. R. Baserga). New York: Marcel Dekker 1971.
30) Skipper, H. E.: Cancer Res. *29*, 2329 (1969).
31) Brooks, P., Lawley, P. D.: Brit. med. Bull. *20*, 91 (1964).
32) Brooks, P. in: Chemotherapy of cancer (ed. P. A. Plattner), p. 32. Amsterdam: Elsevier 1964.
33) Golder, R. M., Martin-Guzman, G., Jones, J., Goldstein, N. O., Rotenberg, S., Rutman, R. J.: Cancer Res. *24*, 964 (1964).
34) Linford, J. H., Froese, A., Israels, L. G.: Nature *197*, 1068 (1963).
35) Riches, P. G., Harrap, K. R.: Cancer Res. *33*, 389 (1973).
36) Wheeler, G. P.: Cancer Res. *23*, 1334 (1963).
37) Bratzel, R. P.: Cancer Chemother. Rep. *26*, 1 (1963).
38) Schmidt, L. M., Fradkin, R., Sullivan, R., Flowers, A.: Cancer Chemother. Rep., Suppl. 2, Parts I, II and III. National Institutes of Health, Bethesda, Maryland: 1965.
39) Boesen, E., Davis, W.: Cytotoxic drugs in the treatment of cancer. London: Arnold 1969.

T. A. Connors

40) Brule, G., Eckhardt, S. J., Hall, T. C., Winkler, A.: Drug therapy of cancer. Geneva: World Health Organisation 1973.
41) Ansfield, F. J.: Chemotherapy of malignant neoplasms. Springfield, Illinois: C. C. Thomas 1973.
42) Whisson, M. E., Connors, T. A.: Nature 206, 689 (1965).
43) Clifford, P. in: Burkitt's lymphoma (eds. D. P. Burkitt and D. H. Wright), p. 52. Edinburgh and London: E. and S. Livingstone 1970.
44) Roberts, J. J. in: DNA repair mechanisms, p. 41. Stuttgart and New York: F. C. Schattauer Verlag 1972.
45) Adamson, R. H.: Ann. N. Y. Acad. Sci. 179, 432 (1971).
46) Calvert, N., Connors, T. A., Ross, W. C. J.: Europ. J. Cancer 4, 627 (1968).
47) Weber, G. in: Liver cancer. I. A. C. R. Scientific Publication No. 1, p. 69. Lyon: 1971.
48) Ross, W. C. J.: Biochem. Pharm. 8, 235 (1961).
49) Connors, T. A., Mitchley, B. C. V., Rosenoer, V. M., Ross, W. C. J.: Biochem. Pharm. 13, 395 (1964).
50) Davis, W., Ross, W. C. J.: J. Med. Chem. 8, 757 (1965).
51) Meyza, J., Cobb, L. M.: Cancer 27, 369 (1971).
52) Bukhari, A., Connors, T. A., Gilsenan, A. M., Ross, W. C. J., Tisdale, M. J., Warwick, G. P., Wilman, D. E. V.: J. Natl. Cancer Inst. 50, 243 (1973).
53) Connors, T. A., Whisson, M. E.: Nature 210, 866 (1966).
54) Connors, T. A., Farmer, P. B., Foster, A. B., Gilsenan, A. M., Jarman, M., Tisdale, M. J.: Biochem. Pharm. In press 1973.
55) Young, C.: Personal communication.
56) Trams, E. G., Nadkarni, M. V.: Cancer Res. 16, 1069 (1956).
57) Connors, T. A.: Biochem. Pharmacol. In press 1974.
58) Connors, T. A., Cox, P. J., Farmer, P. B., Foster, A. B., Jarman, M.: Biochem. Pharm. In press 1973.
59) Brock, N., Arnold, M., Bourseaux, F.: Angew. Chem. 70, 539 (1958).
60) Foley, G. E., Friedman, O. M., Drolet, B. P.: Cancer Res. 21, 57 (1961).
61) Brock, N.: Cancer Chemother. Rep. 51, 315 (1967).
62) Struck, R. F., Kirk, M. C., Mellett, L. B., El Dareer, S., Hill, D. L.: Mol. Pharmacol. 7, 519 (1971).
63) Hill, D. L., Kirk, M. C., Struck, R. F.: J. Amer. Chem. Soc. 92, 3207 (1970).
64) Alarcon, R. A., Meienhofer, J.: Nature, New Biology 233, 250 (1971).
65) Colvin, M., Padgett, C. A., Fenselau, C.: Cancer Res. 33, 915 (1973).
66) Phillips, B. J.: Biochem. Pharm. In press 1973.
67) Takamizawa, A., Matsumoto, S., Iwata, T., Katagiri, K., Tochino, Y., Yamaguchi, K.: J. Amer. Chem. Soc. 95, 985 (1973).
68) Ball, C. R., Connors, T. A.: Biochem. Pharm. 16, 509 (1967).
69) Goldenthal, E. I., Nadkarni, M. V., Smith, P. K.: Radiat. Res. 10, 571 (1959).
70) Calcutt, G., Connors, T. A.: Biochem. Parm. 12, 839 (1963).
71) Ball, C. R.: Rec. Res. Cancer Res. 21, 26 (1969).
72) Ball, C. R., Connors, T. A., Double, J. A., Ujhazy, V., Whisson, M. E.: Int. J. Cancer 1, 319 (1966).
73) Roberts, J. J., Crathorn, A. R., Brent, T. P.: Nature 218, 970 (1968).
74) Gaudin, D., Yielding, K. L.: Proc. Soc. exp. Biol. Med. 131, 1413 (1969).
75) Clarke, D. A., Barclay, R. K., Stock, C. C., Rondestedt, C. S.: Proc. Soc. exp. Biol. Med. 90, 484 (1955).
76) Carter, S. K., Friedman, M. A.: Europ. J. Cancer 8, 85 (1972).
77) Audette, R. C. S., Connors, T. A., Mandel, H. G., Merai, K., Ross, W. C. J.: Biochem. Pharm. 22, 1855 (1973).
78) Preussman, R., Druckrey, H., Ivankovic, S., Hodenberg, A.: Ann. N. Y. Acad. Sci. 163, 697 (1969).
79) Carter, S. K., Schabel, F. M., Broder, L. E., Johnston, T. P.: Adv. Cancer Res. 16, 273 (1972).

80) D'Angelo, J. M., Groth, D. P., Vogler, W. R.: Proc. Int. Cancer Congr. 10th, 1970. Oncology, 1970, Abst. p 408. (Eds Clark, Cumley, McCay and Copeland). Houston: Year Book Medical Publishers.
81) Green, S., Bodansky, O.: Proc. Amer. Assoc. Cancer Res. 8, 23 (1967).
82) Wheeler, G. P., Bowdon, B. J.: Cancer Res. 28, 52 (1968).
83) Rosenberg, B., van Camp, L., Krigas, T.: Nature 205, 698 (1965).
84) Rosenberg, B., van Camp, L., Grimley, E., Thomson, A. J.: J. Biol. Chem. 242, 1347 (1967).
85) Rosenberg, B., van Camp, L., Trosko, J. E., Mansour, V. H.: Nature 222, 385 (1969).
86) Cleare, M. J., Hoeschele, J. D.: Platinum Metals Review 17, 2 (1973).
87) Connors, T. A., Jones, M., Ross, W. C. J., Braddock, P. D., Khokhar, A. R., Tobe, M. L.: Chem. Biol. Interactions 5, 415 (1972).
88) Howle, J. A., Gale, G. R.: Biochem. Pharm. 19, 2757 (1970).
89) Roberts, J. J., Pascoe, J. M.: Nature 235, 282 (1972).
90) Thomson, A.: Rec. Res. Cancer Res. In press.
91) Connors, T. A.: Proc. VIIth International Congress of Chemotherapy. In: Adv. Antimicrobial and Antineoplastic Chemotherapy 2, 237 (1972).
92) Wiltshaw, E.: Rec. Res. Cancer Res. In press.
93) Culvenor, C. C. J., Downing, D. T., Edgar, J. A.: Ann. N. Y. Acad. Sci. 163, 837 (1969).
94) Baker, B. R.: Design of active-site-directed irreversible enzyme inhibitors. New York: J. Wiley & Sons 1967.
95) Stock, J. A.: Chemotherapy of neoplastic diseases. In: Experimental chemotherapy (eds. R. J. Schnitzer and F. Hawking), vol. 5, Part II, p. 333. New York: Academic Press 1967.
96) Miller, E. C., Miller, J. A.: Ann. N. Y. Acad. Sci. 163, 731 (1969).
97) Schoental, R.: Ann. Rev. Pharmacol. 7, 343 (1967).

Received July 16, 1973

Synthetic Interferon Inducers*

Dr. Erik De Clercq**

Rega Institute for Medical Research, University of Leuven, Leuven, Belgium

Contents

* This study was supported by a grant from the Belgian F.G.W.O. (Fonds voor Genees-
kundig Wetenschappelijk Onderzoek).

** 'Aangesteld Navorser' of the Belgian N.F.W.O. (Nationaal Fonds voor Wetenschappelijk
Onderzoek).

E. De Clercq

1. Introduction

Since the original discovery of Isaacs and Lindenmann[1] on interferon induction by heat-inactivated influenza virus in 1957, ample evidence has been provided that not only viruses but many other microorganisms and even microbial extracts, synthetic compounds and immune recognition mechanisms are able of eliciting an interferon response in the appropriate host or cell culture system. During the past years the number of interferon inducers has been growing so rapidly that a review on interferon inducers is nearly outdated at the time it is in print[2]. Interferon and its inducers have been covered in several recent review articles[2-14]. The inducers can be classified schematically in four major groups, as shown in Table 1. Only the third group (synthetic interferon inducers) will be discussed in the present review.

Table 1. General review of interferon inducers[1])

I.	Microorganisms	Viruses, chlamydiae, rickettsiae, bacteria, protozoa
II.	Microbial and Cellular extracts	Double-stranded RNAs, endotoxins, cycloheximide, phytohemagglutinin, concanavalin A
III.	Synthetic compounds	Polycarboxylates, polynucleotides, fluorene derivatives
IV.	Immune recognition	Renewed exposure of sensitized lymphocytes to specific antigen, antilymphocytic serum, mixture of genetically dissimilar lymphocytes

[1]) For details and references on classes I, II and IV, see Refs.[5, 13].

Interferon has originally been defined as *a cellular protein, produced in response to, and acting by preventing the replication of a virus invading the cell*. This definition should be revised in the sense that interferon can also be induced by *non-viral agents* and is equipped with a wide spectrum of activities, including *non-antiviral activities*. Interferon not only interferes with virus replication, but has also been shown to inhibit the replication of microorganisms, phylogenetically higher than viruses (Chlamydiae, Rickettsiae, bacteria, protozoa), as well as the multiplication of normal and transformed cells. Interferon may either enhance ('priming') or suppress ('blocking') the ability of the cells to produce interferon. Interferon has also been shown to stimulate the phagocytic activity of macrophages and the specific cytotoxicity of lymphocytes for target (tumor) cells. Finally, interferon renders the cells more susceptible to the cytotoxicity of double-stranded RNAs. The different aspects of interferon action have been dealt with in a recent review article by De Clercq and Stewart[14] All activities appear to reside in the same molecule, or, at least in a class of physicochemically very similar molecules.

174

Interferon itself has never been obtained in pure form. With the most highly purified interferon preparations specific activities of more than one million units per mg protein have been obtained, yet a very high proportion of the protein in these preparations is inert material (one *unit of interferon* is defined as the reciprocal of the highest dilution of a sample that reduces viral plaque formation in cell cultures by 50%). Although the chemical structure of the interferon molecule has not been determined, it is generally believed to be a protein or glycoprotein. It is species specific, that is active only in the cells of the animal species in which it has been induced. It is nondialyzable, nonsedimentable at 105,000 g for 2 h, stable to acid at pH 2.0 and has electric points close to pH 7.0. Its activity is destroyed by proteolytic enzymes. As elaborated by Carter[15], interferon may be dissociated in subunits. Viral-induced interferon would be a dimer and poly(I) · poly(C)-induced interferon would be an octamer. The oligomers are apparently formed by doubling of the previous unit of aggregation, *i.e.* monomers → dimers → tetramers → octamers. The monomer itself would have a molecular weight of 12,000 (human interferon) or 19,000 (mouse interferon).

2. General Review of All Interferon Inducers and Their Interferon-inducing Characteristics

The diversity of interferon inducers (Table 1) and their interferon-inducing characteristics is so impressive that it is hard to believe that they all operate through the same mechanism. Some inducers, such as viruses and double-stranded RNAs are active *in vivo* (intact animals) and *in vitro* (cell cultures), other inducers (*e.g.* bacteria, endotoxins, polycarboxylates, fluorene derivatives) are active only *in vivo* or in leukocyte cultures (lymphocytes or macrophages). Microorganisms (viruses, Chlamydiae, Rickettsiae, bacteria, protozoa) generally stimulate a late interferon response (peak titer at 8 h or later); extracts thereof (double-stranded RNAs, endotoxins) stimulate an early interferon response (peak titer at 2 h). Polycarboxylates and fluorene derivatives also elicit a late interferon response (peak at 16—24 h). Some agents do not induce interferon, unless the cells (lymphocytes) have been previously sensitized to the agent (*e.g.* diphtheria toxoid, tetanus toxoid, tuberculoprotein[16]). Interferon production by sensitized lymphocytes upon renewed contact with the specific antigen has been alluded to as 'immune recognition'. This specific immunologic reactivity is also operative *in vivo*: Stinebring and Absher[17] and Salvin et al.[18] found large amounts of interferon produced upon injection of tuberculoprotein in mice infected with *Mycobacterium tuberculosis*, whereas uninfected mice did not respond to the injection of tuberculoprotein. Most interferon stimuli are only active upon parenteral (intravenous, intraperitoneal, ...) administration; however, tilorone dihydrochloride (see below) and endotoxin[19] are also effective when given orally.

The consensus that there are two distinct patterns of interferon production *in vivo*[20], the *'endotoxin type'* or release of preformed interferon, and the *'virus type'* or induction of *de novo* synthesis of interferon cannot be longer sustained. This distinction was essentially based on the differential effects of cycloheximide on endotoxin- and virus-induced interferon[21]. When injected simultaneously with or shortly before the inducer, at doses which effectively blocked protein synthesis, cycloheximide enhanced the interferon response to endotoxin, yet reduced the interferon response to virus. In both cases, interferon production was measured 6–10 h after injection of the inducer. It was established later[22], however, that the differential effects observed with cycloheximide did not depend on the type of interferon inducer employed but on the time that cycloheximide was administered during the process of interferon production. For a variety of interferon inducers belonging to either the 'endotoxin' or 'virus type', cycloheximide reduced the interferon response during the initial phase of interferon production but increased interferon production during its declining phase, regardless of the inducer tested[22]. In as far as the action of cycloheximide may be interpreted to block protein synthesis, these results suggest that all interferon inducers require new protein synthesis for both the initiation and the termination of interferon production.

Could there be a common principle explaining the interferon-inducing capacity of all major classes of interferon inducers (microorganisms, microbial and cellular extracts, synthetic compounds, immune recognition)? An interesting, yet highly speculative hypothesis is that all interferon-inducing systems operate through the formation and/or release of double-stranded RNA (ds-RNA). Ds-RNAs have been isolated from mycophages of *Penicillia*, and from other viral sources including reovirus, rice dwarf virus, cytoplasmic polyhedrosis virus and the replicative forms of MS2 and MU9 mutant coliphages. These natural ds-RNAs as well as the synthetic poly(I) · poly(C) are the most potent interferon inducers on a weight basis. There is suggestive evidence to believe that interferon induction by RNA and DNA viruses is mediated by ds-RNA formed during the replicative cycle of the virus, since viral-specific ds-RNAs with interference or interferon-inducing properties have been isolated from cells infected with both RNA (Mengo, influenza) and DNA (vaccinia) viruses[23, 24, 25]. However, the view that virus-induced interferon is invariably based on the formation of an intermediary ds-RNA has not met general acceptance[26–32]: *e.g.* experiments conducted by Gandhi and Burke[30], Gandhi *et al.*[31], and Bakay and Burke[32] failed to demonstrate a correlation between interferon production and ds-RNA synthesis in chick cells infected with either vaccinia virus, adenovirus or UV-irradiated Newcastle disease virus (NDV): in vaccinia virus-infected cells ds-RNA but not interferon was formed, while in adenovirus- or NDV-infected cells, interferon but not ds-RNA was formed. The latter results led Gandhi *et al.*[31] to conclude that the single-stranded RNA of the input virions was responsible for the interferon-inducing capacity of UV-irradiated NDV. Doubts have been raised about this conclusion, since recent experiments of Clavell and Bratt[33] and Meager and Burke[34] indicated that, upon UV-irradiation, NDV loses its infectivity much more rapidly than its RNA synthesizing capacity and interferon-inducing capacity, and that, with large doses of UV-irradiation, the RNA synthesizing capacity and interferon-inducing capacity are lost in parallel. These findings suggest the possibility that the single-stranded RNA of the UV-irradiated virus does not itself induce interferon but does so by serving as a template for the synthesis of, albeit limited amounts of, base-paired RNA.

Double-stranded RNAs have also been isolated from normal, ostensibly uninfected mammalian cells: *e.g.* chick embryo, rat liver, rabbit kidney and HeLa cells[25, 35-38]. The general occurrence of ds-RNA in cells of different animal species and the findings that it can be hybridized to host cell DNA[36] and that its biosynthesis is inhibited by actinomycin D[38] argue against a viral origin of this RNA. These cellular ds-RNAs are capable of inducing interferon in homologous as well as heterologous cells[37, 38]. Although they represent only 0.01 % of the total RNA of the cells[38], it is tempting to speculate that they play some role in the interferon induction process. The role of replicative ds-RNA in the interferon induction process of viruses has already been discussed. Whether other major interferon-inducing systems such as endotoxins, phyto-hemagglutinin, concanavalin A and immune recognition (cf. Table 1) require the formation and/or release of ds-RNA is an entirely open question.

Phytohemagglutinin, concanavalin A and the immune recognition mecha-nisms summarized in Table 1 not only stimulate interferon production in leukocyte cultures but also induce transformation of small lymphocytes to large blast-like cells (blastogenesis). Although the extent of lymphocyte trans-formation is not quantitatively related to the amounts of interferon pro-duced[16, 39, 40], and although a partial separation of the interferon-inducing and blastogenic factors has been achieved in phytohemagglutinin[41], it is not far-fetched to assume that lymphocyte to lymphoblast transformation is a necessary event for interferon production by leukocytes in response to specific or nonspecific mitogens. The blast transformation of lymphocytes is accom-panied by a significant increase in DNA, RNA, protein and phospholipid syn-thesis. The increase in RNA synthesis is one of the earliest and most relevant changes observed in phytohemagglutinin-stimulated lymphocytes[42-44], and whether part of this newly synthesized RNA is base-paired and might serve as the intermediary interferon-inducing principle in the activated lymphocytes is obviously an attractive working hypothesis.

3. Review of the Synthetic Interferon Inducers and Their Interferon-inducing Characteristics

The array of synthetic compounds that are able to induce interferon is as diverse as the general interferon-inducing systems reviewed in Table 1. Included in the class of the synthetic interferon inducers are both high molecular weight (polycarboxylates, polynucleotides, . . .) and low molecular weight (tilorone, . . .) compounds (Table 2). The formula of some of the most representative synthetic interferon inducers are depicted in Figs. 1, 2 and 3.

E. De Clercq

Table 2. Review of the synthetic interferon inducers and their interferon-inducing characteristics

Interferon inducers	Activity of the interferon inducer (*in vitro*: cell cultures; *in vivo*: animals)	Time of appearance (*early*: peak at 2–4 h; *late*: peak at 16–24 h)	References

I. *High molecular weight compounds*

1. *Polycarboxylates*
   - *Pyran* (maleic anhydride divinylether) *copolymer* and other maleic anydride copolymers — *in vivo* (mice, man) / Late / 45–48)
   - *Polyacrylic acid* (PAA) and polymethacrylic acid (PMAA) — *in vivo* (mice) / Late / 48–50)
   - *Polyacetal carboxylic acids:*
     Chlorite-oxidized oxyamylose (COAM)
     Chlorite-oxidized oxyamylopectin (COAP)
     Chlorite-oxidized oxycellulose (COCEL) — *in vivo* (mice) / Late / 51)

2. *Polysulfates*
   Polyvinylsulfate — *in vivo* (mice) / Late / 52)

3. *Polyphosphates*
   a. *Phosphorylated polysaccharides:*
      Phosphomannans, phosphodextrans — *in vivo* (mice, rabbits) / Late/early / 53, 54)
   b. *Polyribonucleotides:* — / Early
      - Homopolymer duplexes: *e.g.*
        poly(I) · poly(C);
        poly(A) · poly(U) — *in vitro/in vivo* (mice, rabbits, man, etc.) / 55–68) and others 62, 69–75)
      - Alternating copolymers: *e.g.*
        poly(I–C); poly(A–U) — *in vitro/in vivo* (mice, rabbits)
      - Homopolymers: *e.g.*
        poly(I) — *in vitro/in vivo* (rabbits) / 58, 61, 76, 77)
      - Complexes of homopolymers with copolymers: *e.g.* poly(I, G) · poly(C);
        poly(I) · poly(C, G) — (*in vitro*)/*in vivo* (mice, rabbits) / 78)
   c. *Polydeoxyribonucleotides:* — / Early
      - Homopolymer duplexes: *e.g.*
        poly(dA) · poly(dT) — *in vitro*/(*in vivo?*) / 53, 57, 62, 70, 74)
      - Alternating copolymers: *e.g.*
        poly d(A–T) — *in vitro*/(*in vivo?*) / 53, 70, 74)
   d. *Polyribonucleotide analogues:* — / Early
      - *Thiophosphate substituted for phosphate:*
        In alternating copolymers: *e.g.*
        poly(s̄I–s̄C); poly(sA–s̄U) — *in vitro/in vivo* (mice, rabbits) / 69, 71, 75)
        In homopolymer duplexes; *e.g.*
        poly(s̄I) · poly(s̄C) — *in vitro/in vivo* (mice, rabbits) / 75, 79)
      - *Vinyl substituted for ribophosphate:*
        In homopolymer duplexes:
        poly(I) · poly(vC) — *in vitro*/(*in vivo?*) / 80)
      - *Fluoro, chloro, azido, methoxy, acetoxy substituted for hydroxy in the 2'-position:*

178

Table 2 (continued)

Interferon inducer	Activity of the interferon inducer (*in vitro*: cell cultures; *in vivo*: animals)	Time of appearance (*early*: peak at 2–4 h; *late*: peak at 16–24 h)	References
In homopolymer duplexes: *e.g.*			
poly(A) · poly(2'-FU)	*in vitro*/(*in vivo*?)		81)
poly(A) · poly(2'-N₃U)	*in vitro*/(*in vivo*?)		82)
poly(2'-OAcA) · poly(U)	*in vitro*/(*in vivo*?)		83)
poly(I) · poly(2'-ClC)	*in vitro*/(*in vivo*?)		79)
poly(I) · poly(2'-OMeC)	*in vitro*/(*in vivo*?)		84)

II. *Low molecular weight compounds*

1. *Tricyclic dialkylaminoalkyl ethers, ketones, sulfonamides, esters*

2,7-Bis(2-diethylaminoethoxy) fluoren-9-one or *tilorone dihydrochloride* and related compounds: *e.g.*	*in vivo* (mice, rats)	Late	85–91)
– fluorenone ethers			
– fluorenone esters			
– fluorene ketones			
– fluorenone ketones			
– dibenzofuran ketones			
– xanthone ethers			
– anthraquinone ethers			
– anthraquinone sulfonamides			
– fluoranthene esters			

2. *Thiazine and acridine derivatives*

– *Thiazine derivatives:*			
Toluidine blue, methylene blue	*in vivo* (mice)	Early-late	92, 93)
– *Acridine derivatives:*			
Acridine orange, trypaflavine	*in vivo* (mice)	Early-late	93)
Acranil, mepacrine	*in vivo* (mice)	Late	94)

3. *N,N-dioctadecyl-N',N'-bis-(2'-hydroxyethyl) propanediamine*	*in vivo* (mice)	Late	95)

Some structural features have been recognized in the high molecular weight compounds (molecular weight exceeding 10,000, regular and dense sequence of negative charges, a stable primary or secondary structure, . . .), to which the compound should adhere in order to be active as an interferon inducer. It has, so far, been impossible to trace such features in the structure of the low molecular weight compounds. The sole characteristic that appears to be common for

E. De Clercq

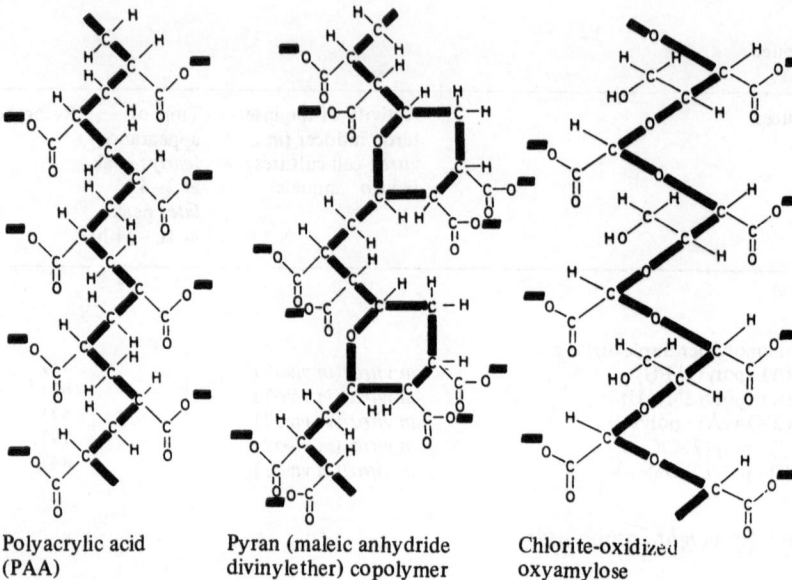

Polyacrylic acid
(PAA)

Pyran (maleic anhydride
divinylether) copolymer

Chlorite-oxidized
oxyamylose
(COAM)

Fig. 1. Polycarboxylates

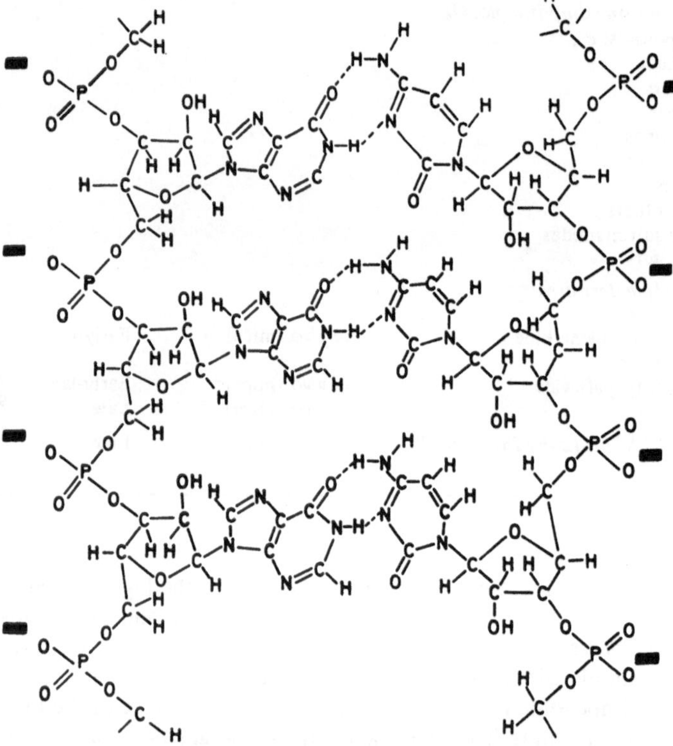

Fig. 2. Polynucleotides [*e.g.* poly(I) · poly(C)]

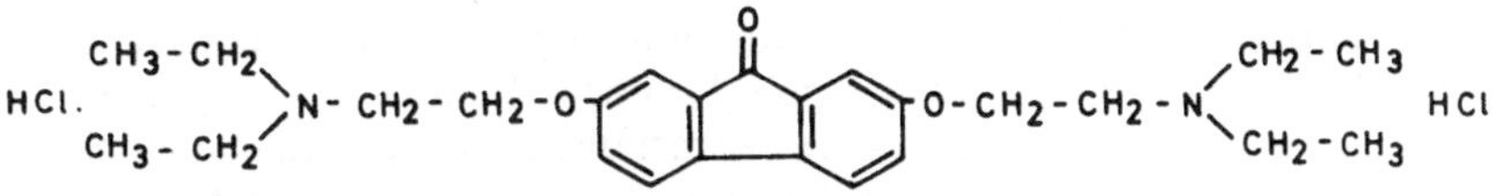

CH_3-CH_2
$HCl.$ $N-CH_2-CH_2-O$ ⬡⬡⬡ $O-CH_2-CH_2-N$ HCl
CH_3-CH_2 CH_2-CH_3 / CH_2-CH_3

Tricyclic dialkylaminoalkyl ethers [e.g. 2,7-bis(2-diethylaminoethoxy) fluoren-9-one dihydrochloride or tilorone dihydrochloride]

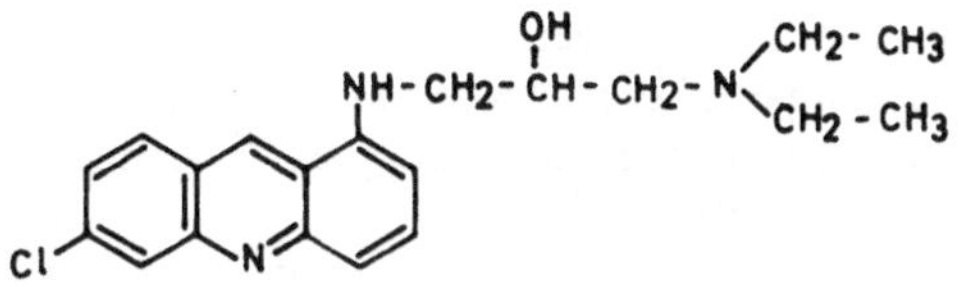

OH
$NH-CH_2-CH-CH_2-N$ CH_2-CH_3 / CH_2-CH_3
Cl ⬡⬡⬡ N

Acridine derivatives (e.g. acranil)

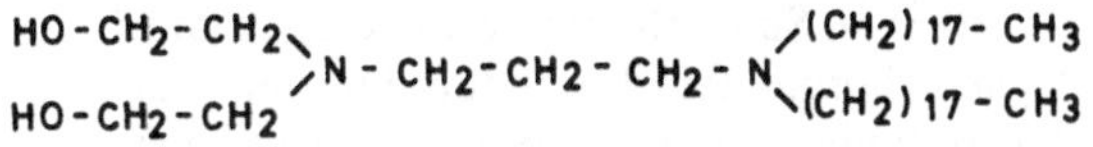

$HO-CH_2-CH_2$
$N-CH_2-CH_2-CH_2-N$ $(CH_2)_{17}-CH_3$ / $(CH_2)_{17}-CH_3$
$HO-CH_2-CH_2$

N,N-dioctadecyl-N',N'-bis(2'-hydroxyethyl)propanediamine

Fig. 3. Low molecular weight interferon inducers

all low molecular weight interferon inducers listed in Table 2, is that they are all *diamines*. Whatever the molecular organization and atomic distances between the two amine groups may be, they do not seem very important.

Tilorone was the first synthetic small molecular weight substance to be described as an orally active interferon inducer[85, 86]. Glutarimide antibiotics[20, 21], such as cycloheximide, acetoxycycloheximide, streptimidone, streptovitacin A, and aminoglycoside antibiotics, such as kanamycin[96] and tenuazonic acid[20] have also been reported to stimulate interferon production, but these effects were obtained with rather high doses of the antibiotics: e.g. the amounts of cycloheximide required for circulating interferon production in the mouse caused an irreversible inhibition of protein synthesis leading to the death of the animal. More recently, basic dyes such as toluidine blue and methylene blue[92, 93] have been reported to induce interferon in mice. However, the latter findings have not been confirmed yet. Some attempts (De Clercq, unpublished data, 1973) to do so, have failed thus far.

It should be pointed out that the interferon inducers presented in Table 2 are widely different in activity, polyribonucleotide duplexes [such as poly(I) · poly(C)] being superior to most other polynucleotides, especially those in which the 2'–OH group has been replaced by a 2'–H, 2'–F, 2'–Cl, 2'–N₃, 2'–O–CH₃ or 2'–O–CO–CH₃ group (see Table 7). The interferon inducers listed in Table 2 also differ in the kinetics of interferon production: polycarboxylates (e.g. pyran copolymer, PAA, PMAA, COAM) and fluorenone

181

derivatives (*e.g.* tilorone) cause a late interferon response (peak titer at 16 to 24 h), whereas polynucleotides [*e.g.* poly(I) · poly(C)] initiate an early interferon response (peak titer at 2 to 4 h). Characteristically, the former are only effective *in vivo* (animals), whereas the latter are able to induce interferon *in vitro* and *in vivo*. All attempts to induce interferon with pyran copolymer, PAA, PMAA, COAM and tilorone in cell cultures have failed[51, 87, 97]. Leukocyte cultures (mouse peritoneal cells) have been reported to produce some interferon in response to pyran copolymer[58], but this finding has not been widely confirmed. Similarly, tilorone has been found to induce interferon in some leukocyte cultures (human peripheral leukocytes[98]), but failed to do so in other leukocyte cultures (human peripheral leukocytes[99], mouse peritoneal lymphocyte and macrophage cell cultures[87, 89]).

A third characteristic difference in the interferon-inducing capacity of polynucleotides [poly(I) · poly(C)] and the other interferon inducers (polycarboxylates, tilorone) is the species dependent responsiveness (Table 3). Polycarboxylates and tilorone are quite effective in mice, yet failed to induce circulating interferon in rabbits[101, 102]. Similarly, polycarboxylates failed to induce interferon in rats[100], and tilorone failed to induce interferon in chicken and man[103, 104]. Polycarboxylates (pyran copolymer) have been shown to stimulate the production of low levels of circulating interferon in man[47]. Most animal species have been found to respond to the interferon-inducing capacity of polynucleotides, although rabbits appeared to be considerably more sensitive than men. The interferon titers obtained in man were been generally lower than those observed in animals.

Table 3. Interferon-inducing capacity (expressed semi-quantitatively) of synthetic interferon inducers in different animal species

Animal species	Polycarboxylates References	Polynucleotides References	Tilorone References
Mouse	++ 45, 46, 48–51)	+++ 59, 64–66, 68, 78)	+++ 85–87, 89–91)
Rat	– 100)	+++ ?	++ 88)
Rabbit	– 101)	++++ 55, 69, 71, 76, 78, 79)	– 102)
Chicken	?	++ 103)	– 103)
Man	+ 47)	+ 67)	– 104)

Our further discussion will be limited to an analysis of the mechanism of interferon induction by
polycarboxylates (such as PAA, pyran copolymer and COAM) (Fig. 1),
polynucleotides [such as poly(I) · poly(C)] (Fig. 2) and
low molecular weight compounds (such as tilorone) (Fig. 3).

Other synthetic compounds, such as polysulfates, phosphomannans, phospho-dextrans, thiazine derivatives, acridine derivatives, and the N,N-dioctadecyl-N',N'-bis(2'-hydroxyethyl) propanediamine will not be further discussed because their interferon-inducing activities have only been reported occasionally and have not been explored in sufficient detail.

4. Structural Requirements for Interferon Induction by Synthetic Interferon Inducers

4.1. Polycarboxylates

The interferon-inducing capacity of polycarboxylates depends on a series of structural characteristics, summarized in Table 4. The first requirement is a sufficiently high molecular weight. The dependence of interferon production and antiviral activity on molecular weight was most thoroughly studied with PAA. De Clercq and De Somer found that both induction of cellular resistance to virus infection *in vitro*, interferon production *in vivo* (mice) and inhibition of vaccinia virus-induced tail lesions *in vivo* (mice) increased with increasing the molecular weight of the polymer from 25,000 to 1.000,000[5, 105]. Niblack[50] demonstrated that PAA had to exceed a critical molecular weight of 1,000 to induce circulating interferon in the mouse, and Mohr *et al.*[106] found similar molecular size requirements for the effects of PAA on both the release of restrictions on nuclear DNA template activity and the prolongation of the life-span of mice inoculated with L1210 leukemia cells.

Table 4. Interferon induction by polycarboxylates:
structural requirements

---

1. *High molecular weight* (for PAA: exceeding 1,000)

2. *Regular and dense sequence of negative charges (carboxyl groups)*

3. *Primary stability*
   (−C−C−C−C−C-backbone: PAA, PMAA, pyran copolymer)
   (−C−C−O−C−O-backbone: COAM, COAP, COCEL)

4. *Spatial configuration (steric orientation of negative charges)*

A second structural feature for interferon induction by polycarboxylates is the presence of negative charges. They may be placed in either alternate (PAA) or adjacent (pyran copolymer) position but should occur in a regular and dense sequence. The polyanionic character of the polymer appears to be a prerequisite for interferon induction, antiviral activity and antibacterial activity, since uncharged polymers such as dextran, polyacrylamide, non-carboxylated polyethylene analogues of pyran copolymers and incompletely oxidized amylose were devoid of any of these activities[48,49,51,107].

A third requirement for interferon induction by polycarboxylates is structural stability. The stability of PAA, PMAA and pyran copolymer resides in their −C−C−C−C-backbone. This makes them poorly metabolized within the organism and may account for their deposition in the reticulo-endothelial cells[47], their toxicity and their prolonged prophylactic activity against some virus infections. Such prolonged prophylactic activity lasting for at least 2 months has been noted with PAA in mice infected intravenously with vaccinia virus[105] as well as with pyran copolymer in mice infected intraperitoneally with Mengo virus[48]. COAM and related oxypolysaccharides (COAP, COCEL) have a −C−C−O−C−O-backbone. This sequence is more readily biodegradable than the −C−C−C−C−C-sequence of the backbone of PAA and pyran copolymer. Accordingly, COAM is less toxic and its prophylactic antiviral activity wears off faster than that of PAA and pyran copolymer[108].

Molecular weight, negative charges and structural stability are not the sole requirements for interferon induction by polycarboxylates, since marked differences have been found in antiviral activity among polycarboxylates with similar molecular weight, anionic character and structural stability. These differences in activity may be attributed to differences in spatial configuration or steric orientation of the negative charges. Such a change in charge distribution may explain the markedly reduced interferon production and antiviral activity of PAA upon implantation of methyl groups[49, 105] (hereby converting PAA into PMAA) and of pyran copolymer upon implantation of benzene rings[48].

4.2. Polynucleotides

The structural requirements for interferon induction by synthetic polynucleotides are summarized in Table 5.

Table 5. Interferon induction by synthetic polynucleotides: structural requirements

---

1. Strand continuity: High molecular weight
2. Base pairing: Highly ordered secondary structure
3. For poly(I) · poly(C): Requirements of strand continuity and base pairing more stringent in the poly(I) strand than in the poly(C) strand
4. Adequate resistance to nucleolytic degradation
5. Presence of 2-hydroxyl groups

4.2.1. Sufficiently High Molecular Weight

Conflicting data have appeared on the influence of molecular size on the interferon-inducing capacity of poly(I) · poly(C). Thus, Lampson et al.[109] found that reduction of the molecular size of poly(I) · poly(C) by sonic radiation led to a gradual decrease in antiviral activity. Shiokawa and Yaoi[110], however, reported enhancement of the interferon-inducing activity upon sonication of poly(I) · poly(C) preparations with sedimentation values larger than 10S. Similarly, Tytell et al.[111], Mohr et al.[106], and Carter et al.[112] found that the antiviral activity of poly(I) · poly(C) depended more upon maintaining a high molecular weight of poly(I) than of poly(C), whereas Wacker et al.[113], Niblack and McCreary[114] and Morahan et al.[115] concluded that poly(I) and poly(C) behaved similarly no matter which was the high or low molecular weight component in the complex. A close analysis of the data reported reveals that they are less conflicting than appears at a first glance. It all depends on the range of molecular sizes studied (Fig. 4). There is a critical range of molecular sizes [spread from approximately 2S (molecular weight 10,000 daltons) to 5S (molecular weight 120,000 daltons)] in which a reduction of the molecular size is accompanied by a nearly linear loss of antiviral activity, and in which maintaining a high molecular size of poly(I) is more important for activity than maintaining a high molecular size of poly(C). The molecular sizes of the poly(I) and poly(C) preparations studied by Tytell et al.[111], Mohr et al.[106] and Carter et al.[112] fell in this range, whereas those studied by Wacker et al.[113] and Morahan et al.[115] fell respectively below the 2S and above the 5S limits. Accordingly, Wacker et al.[113] and Morahan et al.[115] found that the chemical identity of the higher or lower molecular components of the poly(I) · poly(C) duplex did not influence the antiviral activity. The different poly(I) and poly(C) preparations used by Niblack and McCreary[114] fell into the critical range, yet the authors pretended that the chemical identity of the higher or lower molecular weight moieties of the poly(I) · poly(C) complex seemed insignificant. It should be pointed out, however, that the poly(I) and poly(C) fractions used in these experiments were obtained by thermal degradation (125 °C) for various times (5 to 180 min); as acknowledged by the authors, their preparations were highly polydisperse. No estimation of size distribution within samples were made, but, according to the Tm versus molecular weight data, these samples must have contained a large amount of rather small molecules. Thus, the data reported by Niblack and McCreary[114] should be interpreted with caution.

The model of the molecular size — interferon-inducing activity relationship proposed in Fig. 4 can also be applied to poly(I) · poly(C) preparations in which the molecular sizes of both strands have been reduced simultaneously. Both Lampson et al.[109] and Black et al.[75] have shown that interferon production *in vivo* (rabbits) and *in vitro* (rabbit kidney or human skin fibroblast cell cultures) was rather resistant to molecular weight reduction unless the molecular size fell beneath the 5S limit. It has been established by Shiokawa and Yaoi[110] that the interferon-inducing capacity of poly(I) · poly(C) may decrease, if its molecular size becomes too large (> 10S). It can be concluded, therefore, that the optimal size of poly(I) · poly(C) for interferon production is situated in the 5S–10S range.

185

E. De Clercq

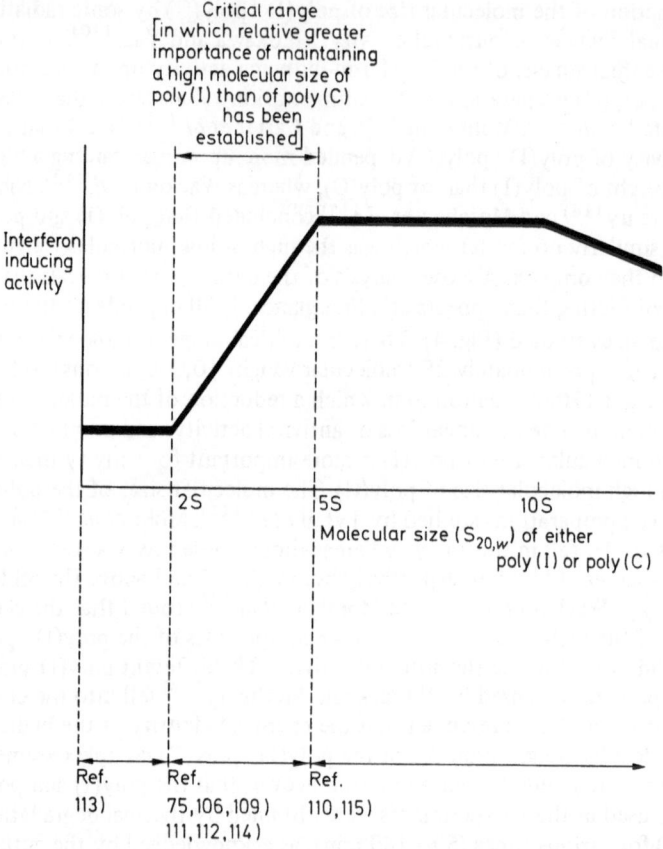

Fig. 4. Influence of molecular size on interferon-inducing activity of poly(I) · poly(C)

4.2.2. Stable, Highly Ordered Secondary Structure

A second structural requirement for the interferon-inducing capacity (antiviral activity) of synthetic polynucleotides is a stable, highly ordered secondary, hence double-stranded structure based on complementary base-pairing. The stability of the complex is reflected in its Tm (thermal stability) value. From a comparative study of different double-stranded RNA duplexes, De Clercq and Merigan[61] and De Clercq et al.[53] concluded that a Tm value higher than 60 °C (calculated for 0.15M Na^+) was needed for full expression of antiviral activity. Since thermal stability represents a valuable measure of the overall stability of double-stranded RNAs[116], double-stranded RNAs with Tm values higher than 60 °C may also be considered to be the most stable ones at 37 °C.

It seems as though both strands of the duplex must stay together in a stable highly ordered form in order to be effective in inducing interferon. The influence of thermal stability on interferon induction is depicted in Fig. 5. This figure may be interpreted as follows: (1) RNA complexes melting out at Tm < 40 °C do not show any activity beyond the activity due to the individual homopolymer components; (2) within the range of 40° to 60 °C, antiviral activity gradually increases with thermal stability; and (3) above 60 °C, no further increase of activity occurs, the antiviral activity having reached its maximal expression. The interferon-inducing activities of the RNA duplexes with *Tm* values above 60 °C may fluctuate around the plateau line: *e.g.* poly(A–U) with a *Tm*: 69 °C is less effective than poly(I) · poly(C) with a *Tm*: 62 °C.

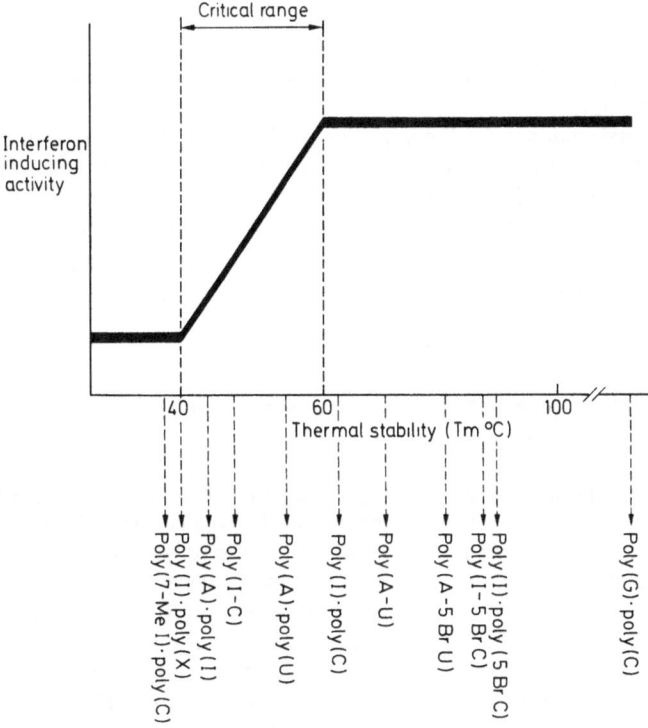

Fig. 5. Influence of thermal stability on interferon-inducing activity of synthetic poly-ribonucleotides

The *Tm*-activity correlation plotted in Fig. 5 does not only accomodate our own data on thermal stability and antiviral activity [53, 61, 69, 71] but also those of others [62, 83, 112, 114, 117, 118]: *e.g.* Niblack and McCreary[114], studying different poly(I) · poly(C) preparations with variations in the molecular size of either component found a nearly linear relationship between antiviral activity, thermal stability (range 46°–60 °C) and molecular size. A similar dependence of antiviral activity on thermal stability was established in a series of oligo(I) · poly(C) complexes by Carter et al.[112]:

oligo(I)$_n$ < 10 · poly(C) (Tm: < 40 °C) was inactive[119],

oligo(I)$_{17}$ · poly(C) (Tm: 47 °C) was marginal,

oligo(I)$_{19}$ · poly(C) (Tm: 50 °C) was definitely active, though 100-fold less than poly(I) · poly(C).

Reduction of the molecular size of poly(C) in poly(I) · poly(C) led also to a decrease in thermal stability and to a concomitant decrease in antiviral activity[112]. However, poly(I) · oligo(C) proved biologically more active than oligo(I) · poly(C), even when they were similar in thermal stability and not greatly different in oligomer chain length. This observation has already been alluded to in the preceding section (influence of molecular size on interferon-inducing activity) and will be further discussed in the following section [differential importance of poly(I) and poly(C) in the antiviral activity of poly(I) · poly(C)].

Additional evidence for the importance of structural stability in the antiviral activity of synthetic polynucleotides was provided by Harnden et al.[118] and Steward et al.[83]. These authors found a decrease in thermal stability and concomitant decrease in antiviral activity of poly(A) · poly(U) and poly(I) · poly(C) upon N-oxidation of poly(A) and poly(C) with m-chloroperbenzoic acid[118] or 2′–O-acetylation of poly(A) with acetic anhydride[83].

4.2.3. Differential Importance of Constituent Homopolymers in Poly(I) · Poly(C)

For interferon induction by poly(I) · poly(C), the structural requirements of strand continuity (high molecular weight) and base-pairing (highly ordered double-stranded structure) are more stringent in the poly(I) strand than in the poly(C) strand (Table 6). In studies reported by Carter et al.[112] two types of structural modifications were imposed to poly(I) · poly(C): strand interruption and mismatching of bases. Both modifications lowered the interferon-inducing activity of the complex. Yet, the decrease was much more significant when the perturbation was imposed upon the poly(I) strand than upon the poly(C) strand: e.g. poly(I$_{39}$,U) · poly(C) and poly(I$_{21}$,U) · poly(C) showed little interferon-inducing activity, while poly(I) · poly(C$_{22}$,U) and poly(I) · poly(C$_{13}$,U) were as active as poly(I) · poly(C) itself. In similar studies, Matsuda et al.[78] found higher interferon production with poly(I) · poly(C$_5$,A) and poly(I) · poly(C$_5$,U) than with poly(I$_5$,A) · poly(C) and poly(I$_5$,U) · poly(C). The unpaired U and A residues in these complexes are expected to be excluded from the helix and to loop out into the solvent, although Wang and Kallenbach[120] purported that the U residues may remain in the helical complex of poly(I) · poly(C$_X$,U) when X is larger than 5.

Table 6. Differential importance of poly(I) and poly(C) in the antiviral activity of poly(I) · poly(C)

Modification	Antiviral activity of poly(I) · poly(C)	References
Reduction in molecular size (e.g. 10S → 2S)		106, 111, 112)
of poly(I) strand	Markedly decreased	
of poly(C) strand	Slightly decreased	
Interruption by unpaired bases (either A, U or G)		78, 112)
in poly(I) strand	Markedly decreased	
in poly(C) strand	Slightly decreased	
Sequential administration		122—124)
poly(C) followed by poly(I)	Decreased	
poly(I) followed by poly(C)	Enhanced	

That the size of the poly(I) moiety is far more critical than the size of poly(C) in determining the antiviral activity of the poly(I) · poly(C) complex has been emphasized in several reports[106, 111, 112, 121]: e.g. Tytell et al.[111] found no significant decrease in antiviral activity unless the average molecular weight of poly(I) was reduced to 190,000 (550 residues) and of poly(C) to 23,000 (70,000 residues). Subsequently, Lampson et al.[121] showed that the molecular size of poly(C) could be reduced to 2.5S (molecular weight 16,000 daltons) without reduction of *in vitro* or *in vivo* protection against virus infections. The differential importance of poly(I) as compared to poly(C) has even been established in the cure of mice inoculated with L1210 leukemia cells: poly(I) · poly(C) complexes were effective in increasing the life-span of these mice whenever the poly(I) was larger than 4.4S and did not depend on the size of poly(C); if the poly(I) was smaller than 4.4S, poly(I) · poly(C) was ineffective regardless of the size of poly(C) with which it was combined[106].

Similarly, thermal stability which may be considered as an important parameter for the antiviral activity of synthetic polyribonucleotides (Fig. 5), was significantly reduced (64.0 → 52.8 °C) if the size of poly(I) was decreased from 13 to 3S, whereas a similar decrease in the size of poly(C) did not markedly affect the Tm of poly(I) · poly(C) (I. Mecs, as referred to by De Clercq and De Somer)[122]. Mohr et al.[106] have confirmed these results.

The relatively greater importance of poly(I) than of poly(C) in the antiviral activity of poly(I) · poly(C) has further been demonstrated in experiments in which the two components were added sequentially to cell cultures[122—124]. Sequential administration of the homopolymers in the order poly(I) followed by poly(C) led to an equal or greater interferon response than obtained after addition of poly(I) · poly(C) itself. However, when the homopolymers were added in the order poly(C) → poly(I) the antiviral activity of the complex was only partially restored. The mechanism by which the order of addition of

189

poly(I) and poly(C) controls interferon induction and antiviral activity has not been completely settled, but, as discussed below, might be explained by a more efficient binding of poly(I) · poly(C) to the cellular receptor sites by its poly(I) strand than by its poly(C) strand[124].

4.2.4. Adequate Resistance to Nucleolytic Degradation

Several methods have been developed to increase the interferon-inducing capacity of polynucleotides.

(1) Addition of polybasic substances such as DEAE-dextran or polylysine has been shown to increase the interferon response to poly(I) · poly(C), or to one of its homopolymer constituents, *in vitro*[57, 60, 62, 63, 73, 77, 112, 125−133] and *in vivo*[134−137]. In neither study conclusive evidence was provided for the mechanism of the stimulatory effect of polycations on interferon induction by poly(I) · poly(C): some authors favored the hypothesis that the stimulatory effect of DEAE-dextran was primarily due to an effect on the cells[128, 129, 131]. Pitha and Carter[130], however, found that the stimulation of interferon production by DEAE-dextran was maximal when a poly(I) · poly(C)/DEAE-dextran complex was formed with a 1:1 P/N ratio, suggesting that the stimulatory effect of DEAE-dextran was primarily due to complex formation with the polynucleotide. Such complex formation may also offer a reasonable explanation for the potentiation of interferon production observed *in vivo*[134−137]. The increased interferon-inducing capacity of the poly(I) · poly(C)/DEAE-dextran was accompanied by an increased resistance to degradation by endonucleases (pancreatic ribonuclease)[127, 130].

(2) Substitution of thiophosphate for phosphate, and

(3) preincubation of the polymer at 37 °C in cell culture medium brought about a parallel increase in the interferon-inducing capacity of the alternating copolymers poly(A−U) and poly(I−C) and their resistance to degradation by endonucleases[69−72]. Potentiation of the interferon response was demonstrated *in vitro* and *in vivo* with the thiophosphate-substituted polynucleotides, but only *in vitro* with the preheated polymers.

Two other systems have been described to enhance interferon production by cells exposed to poly(I) · poly(C):

(4) addition of metabolic inhibitors (cycloheximide, actinomycin D)[138−146] and

(5) pretreatment of the cells with interferon[145−153] or small doses of poly(I) · poly(C)[154]. The former has been referred to as 'superinduction', the latter as 'priming'.

Systems (4) and (5) have been applied successfully to potentiate human interferon production in diploid cell cultures *in vitro*[143−146]. Superinduction may also be operative *in vivo*[22, 59]. Whether priming is operative *in vivo*, has not been assessed yet. The superinduction phenomenon will be discussed below. Neither superinduction nor priming[153] appear to act through increasing the resistance of poly(I) · poly(C) to nucleolytic degradation.

That resistance to nucleolytic degradation may play an important part in the interferon-inducing capacity of poly(I) · poly(C) and other double-stranded polyribonucleotides is suggested by the aforementioned findings pointing to a parallel increase of interferon production and resistance to nucleases upon
  (1) complex formation with DEAE-dextran,
  (2) substitution of thiophosphate for phosphate, and
  (3) thermal activation.
The contention of the importance of resistance to nucleases is further strengthened by the experiments of Nordlund et al.[155] and Stern[156] showing the presence in various animal sera of an enzymatic activity specifically directed against base-paired RNA. This serum enzyme may be responsible for the decreased interferon response to poly(I) · poly(C) in animals treated with heterologous serum before injection of the polynucleotide[157]. It may also account for the finite half-life of ds-RNAs delivered exogenously or endogenously into the blood stream of all animal species.

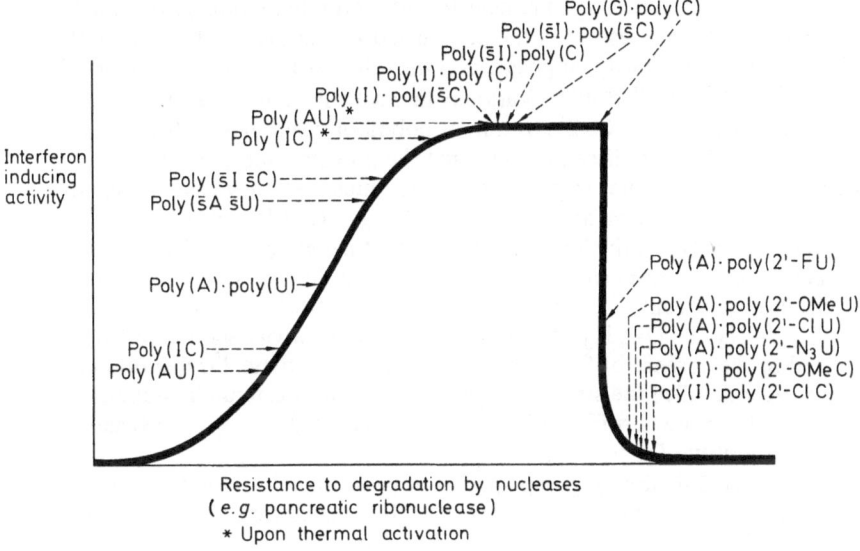

Fig. 6. Influence of resistance to nucleolytic degradation on interferon-inducing activity of synthetic polynucleotides

Although Colby and Chamberlin[62] failed to link the efficiency of polynucleotide interferon inducers to their sensitivity to pancreatic ribonuclease, it is unquestionable that the more effective interferon inducers [poly(I) · poly(C), poly(G) · poly(C), thiophosphate-substituted poly(A–U) and poly(I–C), and thermally activated poly(A–U) and poly(I–C)] are all more resistant to nucleolytic degradation than the lesser active compounds [poly(A) · poly(U), non-activated poly(A–U) and poly(I–C)] (Fig. 6; Refs.[62, 69, 70–72, 75]). However, the polynucleotide duplexes, poly(A) · poly(U), poly(I) · poly(C), in which the 2'-hydroxyl of the pyrimidine strand has been replaced by a 2'-fluoro, 2'-O-methyl, 2'-chloro, or 2'-azido group, are considerably more resistant toward nucleases than their parent compounds[158−163], yet they do not exhibit more than marginal activity[79, 81, 82, 84]. Does it mean that the interferon-inducing activity of polynucleotides is compatible only with a certain range of resistance to nucleolytic degradation, and that at least partial degradation of the polynucleotide is needed for the interferon induction process? The answer to this question cannot be ascertained without further experimentation.

4.2.5. Presence of 2'-Hydroxyl Groups

The presence of 2'-hydroxyl groups is certainly one of the most, if not the most stringent structural requirement for interferon induction by polynucleotides. That the sugar residue in the polynucleotide should be ribose and not desoxyribose was already emphasized in the initial studies of Vilcek et al.[57] and Colby and Chamberlin[62]. Various other substitutions have since then been carried out at the C−2' position of polyribonucleotides (Table 7), and, although all these modifications rendered the polymers more resistant to degradation by nucleases (e.g. pancreatic ribonuclease), they caused a significant decrease in antiviral activity. For the $2'-O-CO-CH_3$ substituted poly(A) · poly(U)[83], the loss in activity should merely be regarded as a consequence of non-compliance with the thermal stability requirement [see note [3] in the legend to Table 7], but, for the other 2'-substituted duplexes ($2'-F$, $2'-Cl$, $2'-N_3$, $2'-O-CH_3$), the loss or decrease in activity should directly be related to the absence of free 2'-hydroxyl groups. All these 2'-substituted polynucleotides fulfilled the structural requirements for interferon induction, as outlined above, viz. a sufficiently high molecular weight ($S_{20,w} > 5S$) and thermal stability (Tm °C $\geqslant 60°$).

Thus, the 2'-hydroxyl group appears to be intimately involved in the interferon induction process of polynucleotides. It is mandatory to find out whether the 2'−OH as such serves as the trigger for the induction of interferon or whether its presence results in a particular steric configuration of the molecule that is recognized by the interferon triggering site. Due to the presence of 2'−OH groups, ds−RNAs are able to form hydrogen bonds with surrounding water molecules, RNA helices (interhelical hydrogen bonds)[165], and other molecules. Ds−RNAs lack this capacity. The $2'-F$, $2'-Cl$, $2'-N_3$ and $2'-O-CH_3$ substituted analogues are also unable to make such hydrogen bonds and, therefore, resemble ds−DNAs.

The differences in interferon-inducing activity between ds−RNAs and ds−DNAs could be ascribed to differences in *configuration* (A form *versus* B form). In the B form (ds−DNA in solution), the planes of the base pairs are almost perpendicular to the helical axis; in the A form (ds−DNA in dehydrated

Table 7. Influence of various 2'-OH substitutions on the interferon-inducing activity of synthetic polyribonucleotide duplexes

Duplex	Substitution[1] of 2'-OH by	Strand modified	Molecular size ($S_{20,w}$) Purine strand	Pyrimidine strand	Thermal stability (Tm °C) 0.1 M Na⁺	0.15 M Na⁺	Resistance of the modified strand to degradation by nucleases	Interferon-inducing activity of the duplex	References
Poly(A) · poly(U)	—	—	7.6–9.2	3.5–4.8–8.0	57°–58°	59°	—	—	79, 81, 84, 163)
	2'-F	Pyrimidine	9.2	12.3	75°	—	Increased	Decreased 10 x	81, 158)
	2'-Cl	Pyrimidine	7.6	7.1	56°	—	Increased	Decreased 100 x	79, 162)
	2'-O-CH₃	Pyrimidine	9.2	9.0	—	70.5°	Increased	Decreased > 100 x	84, 160, 161)
	2'-N₃	Pyrimidine	—[2]	8.0	59°	—	Increased	Decreased > 1000 x	82, 163)
	2'-O-CO-CH₃	Purine	—[2]	—	—	30°[3]	Increased	Decreased > 100 x	83)
Poly(I) · poly(C)	—	—	7.9–9.1	6.0–13.2	62°	64°	—[4]	—	79, 84)
	2'-H	Purine	9.1–12.5	6–12	35°	—	Increased	Decreased up to > 10,000 x	57, 62, 70, 79, 84)
	2'-Cl	Pyrimidine	7.9	8.7	52°	—	Increased	Decreased > 10,000 x	79)
	2'-O-CH₃	Pyrimidine	9.1	9.0	—	61.5°	Increased	Decreased 10,000 x	84, 159, 161)

1) Poly(U) with 2'-OH substituted by 2'-NH₂, and poly(I) and poly(C) with 2'-OH substituted by 2'-O-CO-CH₃ have not been incorporated, because they failed to form regular duplexes with the complementary homopolynucleotides[83, 164]. The 2'-NH₂ and 2'-O-CO-CH₃ substituted homopolymers were devoid of any significant antiviral activity[79, 83].

2) Unchanged as compared to the unsubstituted purine strand[83].

3) Upon 98% acetylation of purine strand; Tm °C was 44° upon 54% acetylation, 53° upon 30% acetylation, and 63° upon 0% acetylation of the purine strand[83]. Since substitution of 2'-O-CO-CH₃ for 2'-OH in the poly(A) component of poly(A) · poly(U) resulted in a significant decrease of the thermal stability of the duplex, the decrease in interferon-inducing activity brought about by the substitution should merely be regarded as a consequence of the decrease in thermal stability.

4) Rather decreased according to Ref.[62].

environments; ds–RNA), they are tilted about 20°. DNA–RNA hybrids exist in the A form[165], yet are quite inactive as interferon inducers[57, 62, 167]. To account for these apparent discrepancies, Kleinschmidt[8] formulated an attractive hypothesis: the activity of the inducer molecule may not primarily depend on its initial structure but on its capability to alter its conformation which would permit it to interact with the cellular receptor site for interferon induction. The superior activity of ds–RNAs may actually reside in their property to undergo internal structural rearrangements [A form (11 residues per turn), A′ form (12 residues per turn), A″ form (11.3 residues per turn)][168, 169]. This greater degree of freedom would permit a more ready adaptation to the conformation of the receptor molecule. DNA–RNA hybrids cannot move out of their tilted base A form[166], and would, therefore, be unable to assume compatibility with the receptor molecule. Whether the decrease in antiviral activity observed with $2'-F$, $2'-Cl$, $2'-N_3$ and $2'-O-CH_3$ substituted polyribonucleotides is also accounted for by a decreased capability to assume such transitional conformations, can only be ascertained by further study (X-ray diffraction).

It is surprising that apparently small changes in only one strand [poly(C) of the poly(I) · poly(C) duplex (*e.g.* substitution of $2'-OH$ by $2'-Cl$, $2'-O-CH_3$ or $2'-H$)] led to such a dramatic decrease in antiviral activity[62, 79, 84], whereas more drastic modifications such as substitution of poly(C) by polyvinyl C[80], and interruption of the continuity of the poly(C) strand by bond breakage[106, 111, 112] or unpaired bases[78, 112] only caused a slight decrease in the antiviral activity of the duplex. The latter observations have been interpreted to mean that in the process of interferon induction by poly(I) · poly(C), poly(I) is relatively more important than poly(C). It has also been demonstrated[124] that the poly(I) · poly(C) duplex is most tightly and most efficiently bound to the cell (surface) if the homopolymers are added in the sequence poly(I) followed by poly(C), suggesting that the duplex is attached to the postulated receptor site by its poly(I) rather than by its poly(C) strand. How can it then be explained that subtle modifications in the poly(C) strand (substitution of $2'-OH$ by $2'-Cl$, $2'-H$, . . .) prohibit or switch off this interaction, whereas gross changes in the poly(C) strand (substitution of ribophosphate by vinyl) do not affect it? Apparently not by assuming that the cellular receptor site for induction of interferon specifically recognizes the $2'-OH$ radicals in the poly(I) strand. A more likely hypothesis is that the $2'-OH$ substituted polynucleotides lack the particular steric configuration, or, in view of the structural transition theory of Arnott[168], the structural flexibility to interact adequately with the receptor site for interferon induction.

4.3. Low Molecular Weight Compounds

The structural requirements governing induction of interferon by the fluorene derivatives (tilorone) and the other low molecular weight interferon inducers can hardly be defined. Varying activities have been obtained with a set of 12 structurally related compounds representing 9 different chemical series:

fluorenone ethers, xanthone ethers, anthraquinone ethers, fluorene ketones, fluorenone ketones, dibenzofuran ketones, anthraquinone sulfonamides, fluorenone esters and fluoranthene esters[91].

The protective activity afforded by these compounds against some experimental virus infections in mice did not always correlate with the circulating interferon titers produced. The ethers and ketones referred to above appeared to be active interferon inducers[91]; the anthraquinone sulfonamide, fluorenone ester and fluoranthene ester tested were inactive[92]. The findings reported so far on the interferon-inducing capacity of low molecular weight compounds do not allow to decide which molecular groups determine their activity.

However, most small molecular weight interferon inducers are diamines, thus bis-basic substances, and, obviously, capable of interacting with nucleic acids. Tilorone has been shown to increase the thermal stability of DNA; its own absorption was depressed and red shifted in the presence of DNA[170]. In addition, tilorone increased the viscosity of DNA and decreased its sedimentation rate, two findings which suggest that tilorone binds to DNA by intercalation[171]. The interferon-inducing capacity of basic dyes such as toluidine blue has been tentatively attributed to an interaction with host cell nucleic acids[92, 93]. If the interferon-inducing capacity of tilorone and other basic substances is really due to an interaction with host cell nucleic acid, *e.g.* a conversion of inert host cell nucleic acid to an active interferon inducer, tilorone and its congeners should be particularly effective interferon inducers in rabbits, for rabbits are extremely sensitive to the interferon-inducing properties of polynucleotides[55, 69, 71, 76, 78, 79, 172]. Unfortunately, rabbits do not respond to tilorone[102], and whether they respond to basic dyes has not been assessed.

5. Mechanism of Interferon Production by Cells Exposed to Synthetic Interferon Inducers

Among the synthetic interferon inducers described (polycarboxylates, polynucleotides, low molecular weight compounds), polynucleotides and more specifically double-stranded polyribonucleotides such as poly(I) · poly(C) are efficient interferon inducers *in vitro* (cell cultures). Therefore, studies on the mechanism of interferon production have almost exclusively been performed with ds–RNAs (or viruses).

It has gained general acceptance that *interferon production is under genetic control*. The mechanism of genetic regulation of interferon production may be very similar to the Jacob-Monod model[173] for genetic regulation of protein synthesis in bacteria. That interferon may be produced through a derepression of the gene coding for the interferon protein was first suggested by Kleinschmidt *et al.*[174]. The suggestion stemmed from their studies with statolon, a fermentation product of *Penicillium stoloniferum*: the active principle was initially identified as an anionic polysaccharide, which eventually proved to be an error[175, 176]: indeed, the interferon-inducing activity of statolon appeared

Hypothetical derepression models of interferon production

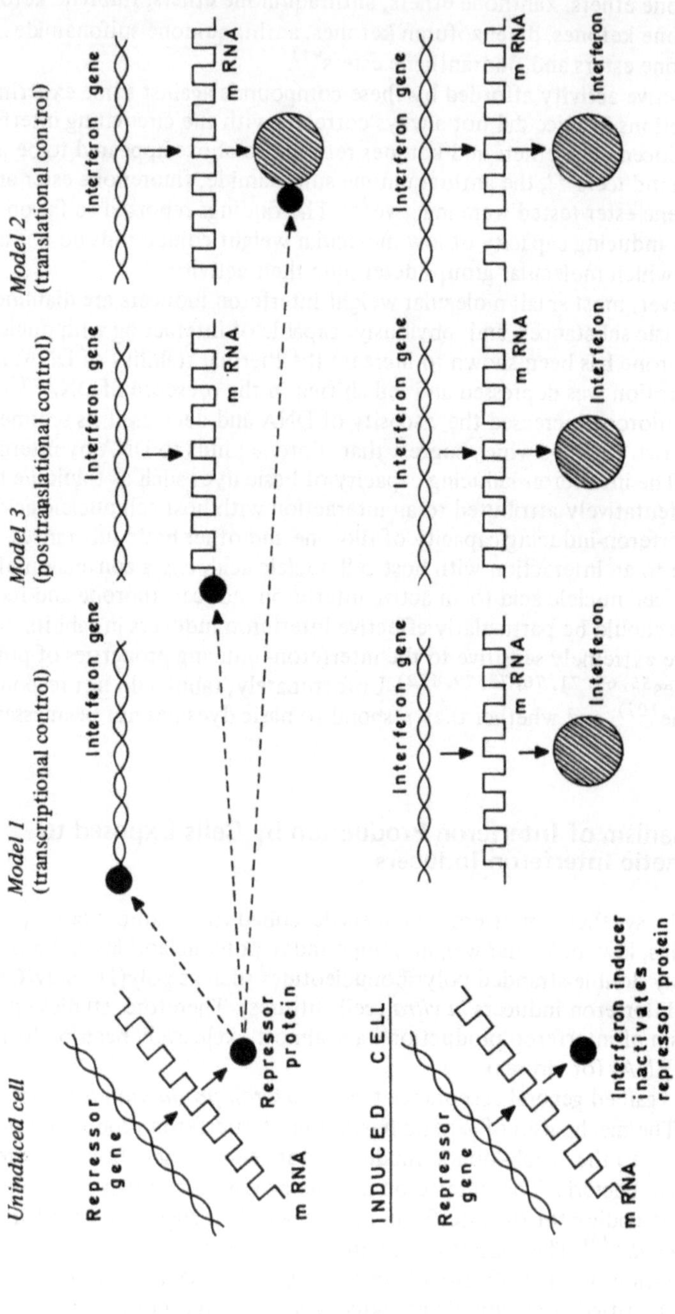

Fig. 7

to reside in the ds–RNA of the mycophages (virus particles derived from the mycelia of the mold).

Several models, yet relatively little direct evidence has been presented to explain the genetic regulation of interferon production. No matter what model may be built to accomodate all known facts, it should be composed, in its simplest form, of an operon (interferon gene), a repressor protein, and a mRNA. The repressor protein may combine with the operator of the operon (referred to as transcriptional control in Fig. 7); in this state, the interferon gene would be repressed. It would be derepressed if the repressor is inactivated, *e.g.* following contact of the cell with the interferon inducer. The derepressed interferon gene would then be transcribed to mRNA, which in turn is translated to interferon. Hypothetically, the repressor protein may also combine with the mRNA for interferon (referred to as translational control), and even with the interferon molecule itself (referred to as posttranslational control in Fig. 7).

Is there direct evidence for the genetic control of interferon production? Two genetic elements that should be involved in interferon production have recently been identified: first, Cassingena *et al.*[177], analyzing the karyotype of a series of monkey-mouse hybrid cell lines, were able to locate the specific genes that code for interferon production, and, second, De Maeyer-Guignard *et al.*[178] demonstrated the presence of interferon mRNA in an RNA extract from cells induced to make interferon. Sofar, no direct evidence has been provided for the existence of other elements in the genetic control of interferon production, viz. the triggering molecule (or derepressor) and the repressor protein. It has even not been established yet whether the ultimate derepressor is the interferon inducer itself or an intermediary substance induced in the cell upon contact with the interferon inducer. Likewise it is not clear whether the repressor protein may be assigned the role of receptor molecule alluded to above.

5.1. Mechanism of Interferon Induction

The existence of a specific cellular receptor site, with which the interferon inducer would combine, was initially postulated by Colby and Chamberlin[62] to explain the high degree of specificity in the induction of interferon by polynucleotides. Allegedly[62], the receptor site would be able to recognize a particular molecular configuration in the structure of polynucleotides. How would this fit with the structural requirements discussed in the preceding section? These requirements may influence the process of interferon induction by polynucleotides in two ways: first, by promoting their delivery, in intact state, at the cellular receptor site, and second, by regulating their interaction with this receptor site. Some requirements (*e.g.* molecular size, thermal stability, adequate resistance to nucleolytic degradation) may be particularly implicated in the delivery step, other requirements (*e.g.* presence of 2-hydroxyl groups) may be directly involved in the recognition step.

That the efficiency of poly(I) · poly(C) as an interferon inducer does not depend on its rate of uptake by the cells has been established in several studies. Thus, De Clercq et al.[74] did not find a significant correlation between the antiviral activity and rate of cell-binding in human diploid cells with eight different polyribo- or polydeoxyribonucleotides of widely varying degrees of antiviral activity. Likewise, poly(I) · poly(C) exposed to a series of eight cell culture lines which differed considerably in sensitivity to the antiviral activity of the polynucleotide bound equally well to the most and least sensitive of these cell cultures[179]. Colby and Chamberlin[62], Black et al.[79], and Field et al.[180] also failed to demonstrate a preferentially increased cellular uptake of poly(I) · poly(C) as compared to other, inactive, or lesser active interferon inducers [poly(I) · poly(dC), poly(I) · poly(2'–ClC), poly(I) and poly(C)]. Although poly(I) · poly(C) was taken up equally well by all sorts of cells, cell-bound poly(I) · poly(C) appeared to be most accessible to extraneous ribonuclease treatment in those cells that were most sensitive to the antiviral activity of the polynucleotide[179]. Thus persistence of the polynucleotide at the outer cell membrane [in as far as sensitivity of cell-bound poly(I) · poly(C) to extraneous nuclease treatment can be considered as a valuable parameter of persistence of the polymer at the cell surface][179] may somehow be related to its antiviral activity.

Does it imply that the triggering or receptor site for interferon induction is located at the cell surface? Not necessarily. To evaluate the possibility that poly(I) · poly(C) may trigger the interferon response from the outer cell membrane, studies have been initiated with poly(I) · poly(C) attached to heterologous cells or red blood cells[181, 182], or solid matrices (sepharose, cellulose, sephadex, cellophane)[183, 184]. Interferon production and antiviral activity[a] were measured with these inducer systems, and it was found that such anchored poly(I) · poly(C) conveyed significant antiviral protection and stimulated interferon production. Although some release of poly(I) · poly(C) from the carrier into the supernatant fluid was observed in all systems studied, the amounts of polynucleotide released were too small to account for the interferon titers produced[181, 182]. However, with poly(I) · poly(C) bound to red blood cells, a direct transfer of polynucleotide of the red blood cell to the interferon-producing (rabbit kidney) cell was noted[182], and, in this particular case, the amounts of polynucleotide taken up by the receptor cells proved large enough to account for the interferon titers produced[182]. These results indicate that mere contact of the polynucleotide with the outer cell membrane may be insufficient for eliciting the interferon response, and, that a firm association with the cell (penetration into the cell?) is required for triggering the induction process. This requirement does, of course, not rule out the possibility that the postulated receptor for interferon production is located at a superficial cell site.

---

a) That the antiviral activity or cellular resistance to virus infection induced by poly(I) · poly(C) and other ds–RNAs in vitro, is due to interferon production has been ascertained in different studies[73, 185–187]. In some peculiar conditions [e.g. poly(I) · poly(C) added at the time of virus inoculation], poly(I) · poly(C) might directly interfere with virus replication by binding to the virus-specific RNA polymerase[188].

5.2. Mechanism of Interferon Synthesis

It has been recognized quite a long time ago that interferon synthesis is achieved by a derepression mechanism. This conclusion was mainly based on the observations that interferon synthesis was blocked in cells treated with inhibitors of RNA and protein synthesis (actinomycin D, puromycin, cyclo-heximide, p-fluorophenylalanine)[189]. More recently, however, Vilcek [138, 139] reported that, depending on the time of administration, actinomycin D and cycloheximide either inhibited or accentuated interferon synthesis: a marked increase in interferon production was noted in cells incubated with poly(I) · poly(C) and then maintained in the continuous presence of cycloheximide as well as in cells treated with actinomycin D at several hours after their exposure to poly(I) · poly(C). These results were originally obtained in rabbit kidney cells but later extended to include other metabolic inhibitors (*e.g.* emetine, pactamycin), other interferon inducers (*e.g.* UV-irradiated NDV) and other cells (*e.g.* human diploid cell cultures)[140–146]. The stimulatory effects of metabolic inhibitors on interferon production can most likely be explained by the inhibition of a cellular regulatory (repressor) protein that controls interferon synthesis[139]. As pointed out above, this repressor protein may do so at either the transcriptional, translational, or posttranslational level (Fig. 7). In more recent studies, in which addition of the metabolic inhibitors to the cell cultures was carefully timed[140, 142], it was suggested that the repressor protein interacted with the interferon mRNA and thereby prevented its trans-lation (translational or, at least, posttranscriptional control of interferon synthesis). Analysis of one summary set of data of Vilcek and Ng[140] is in-structive in this regard: rabbit kidney cells were exposed to poly(I) · poly(C) for 1 h, then incubated with cycloheximide for 3 h, and eventually treated with actinomycin D for 30 min (from 4 to 4.5 h after exposure to the inducer); the cultures were then replenished with inhibitor-free medium, and interferon yields were measured at 22 h: interferon titers were boosted up to about 50,000 units with the treatment regimen described, as compared to *circa* 10,000 units if actinomycin D was omitted, and *circa* 1,000 units if both actinomycin D and cycloheximide were omitted. A plausible explanation for these data is the following. After contact of the cells with poly(I) · poly(C), interferon mRNA is transcribed from the interferon gene. This transcription may be relatively rapid and nearly complete within 1 h after exposure of the cells to poly(I) · poly(C)[141]. The mRNA of repressor, however, would not be transcribed until 4.5 h after addition of the inducer. If this transcription is blocked by actinomycin D added at 4 to 4.5 h, repressor protein cannot be synthesized, and, with interferon mRNA coded for by this time, interferon synthesis cannot be blocked, and it can accumulate to the extraordinary titers mentioned before. It may be concluded, therefore, that the repressor acts only after the interferon mRNA has been synthesized, that is at a posttranscriptional level. Such a posttranscriptional control mechanism has also been proposed by Tomkins *et al.*[190] to explain the effects of actinomycin D on the induction of tyrosine aminotransferase by steroid hormones in hepatoma cells.

6. Biological Activities of Synthetic Interferon Inducers

The biological activity of polycarboxylates, polynucleotides and small molecular weight interferon inducers (*e.g.* tilorone) is not restricted to interferon production. *In vivo*, they exert many other effects on host defense mechanisms, including an increase of the body temperature (pyrogenicity), a stimulation of humoral and cellular immune responses (immuno-adjuvant effect), and an enhancement of the reticulo-endothelial cell activity (phagocytosis) (Fig. 8). In addition, polycarboxylates, polynucleotides and tilorone not only afford protection against virus infections, but also exhibit an inhibitory effect on tumor growth, bacterial infections, fungal infections and protozoal infections (Fig. 8). The different aspects of the *in vivo* biological activities of interferon inducers have been reviewed previously[4, 13] and the reader is referred to these reviews for more details. Taking into account the wide variety of activities which have been associated with interferon[14] (Fig. 8), could interferon production alone account for the various antiviral, antitumoral, antibacterial, antifungal and antiprotozoal activities of the synthetic interferon inducers? It surely could. It is more likely, however, that other host defense mechanisms stimulated by the interferon inducers (such as pyrogenicity, immuno-adjuvant effect and enhanced reticuloendothelial activity) help interferon in achieving its task.

There are three major obstacles that may limit the usefulness of polycarboxylates, polynucleotides and the tilorone congeners in clinical medicine: toxicity, hyporeactivity and immunogenicity. The problem of immunogenicity has been most thoroughly investigated with ds–RNAs such as poly(I) · poly(C). Injected parenterally into rabbits, poly(I) · poly(C) and poly(A) · poly(U) elicited the production of antibodies reactive with the ds–RNA[191–195]. Rabbits with high antibody titers to poly(I) · poly(C) showed also a reduced serum interferon production in response to induction by poly(I) · poly(C)[194, 195]. Similar suppression of the interferon response to poly(I) · poly(C) was reported in New Zealand mice immunized with the polynucleotide[196]. Characteristically, antisera prepared from rabbits injected with poly(I) · poly(C) reacted only with ds–RNAs and not with single-stranded RNAs or with DNA. The one exception was a variable reactivity with poly(I)[195]. The relatively greater reactivity of anti-poly(I) · poly(C) serum with poly(I) as compared to poly(C) may be another example of the relatively greater importance of poly(I) than of poly(C) in the biological activity of the poly(I) · poly(C) duplex (see Section 4.2.3). In preliminary trials in human subjects, Field *et al.*[67] failed to demonstrate the production of anti-poly(I) · poly(C) antibodies, even with relatively large amounts of the polymer administered in repeated doses by the intravenous route. Yet, caution should be advocated in the administration of poly(I) · poly(C) to immunologically hyperreactive hosts (prone to autoimmune disorders: *e.g.* patients with systemic lupus erythematosus, rheumatoid arthritis, . . .), for poly(I) · poly(C) has been shown to accelerate the onset of autoimmune disease in such hosts (New Zealand mice)[197].

Biological activities of synthetic interferon inducers in vivo

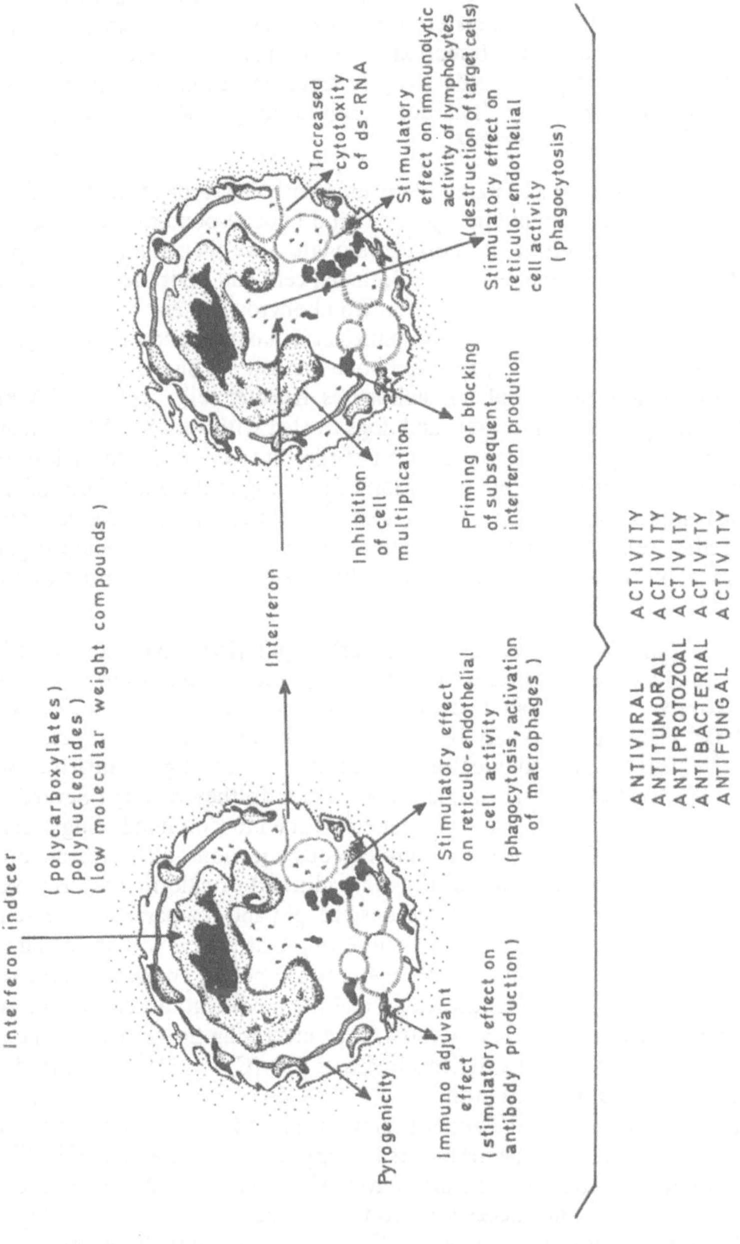

Interferon inducer

(polycarboxylates)
(polynucleotides)
(low molecular weight compounds)

Interferon

Increased
cytotoxity
of ds-RNA

Stimulatory
effect on immunolytic
activity of lymphocytes
(destruction of target cells)

Stimulatory effect on
reticulo-endothelial
cell activity
(phagocytosis)

Inhibition
of cell
multiplication

Priming or blocking
of subsequent
interferon prodution

Interferon

Stimulatory effect
on reticulo-endothelial
cell activity
(phagocytosis, activation
of macrophages)

Immuno adjuvant
effect
(stimulatory effect on
antibody production)

Pyrogenicity

ANTIVIRAL   ACTIVITY
ANTITUMORAL   ACTIVITY
ANTIPROTOZOAL ACTIVITY
ANTIBACTERIAL ACTIVITY
ANTIFUNGAL   ACTIVITY

Fig. 8

Another obstacle confronting the potential use of interferon inducers as chemotherapeutic agents is the development of a state of hyporeactivity, also called tolerance or hyporesponsiveness, following repeated administration of the inducer. The titers of interferon produced as well as the extent of protection conferred against viral infection[64, 68], are markedly reduced during this hyporeactivity period. Hyporeactivity to repeated administration has been demonstrated with nearly all inducers: polycarboxylates[48, 49], polynucleotides[65, 66], tilorone[87, 89, 91].

A consistently emerging problem apparently inherent to the use of interferon inducers, even in animal species that do not respond by circulating interferon production, is *toxicity*. Polycarboxylates owe their toxicity to deposition of the compound in the reticulo-endothelial cells and poor biodegradability. In man, polycarboxylates cause fever, trombocytopenia and basophilic inclusion bodies in the peripheral polymorphonuclear leukocytes and macrophages and in the histiocytes of liver, spleen and bone marrow[47]. Similar basophilic granules were found in the peripheral leukocytes, Kupffer cells, spleen and lymph node macrophages in mice, rats and dogs treated with tilorone[198]. Humans subjected to topical ocular application of tilorone were troubled by blurred vision apparently caused by deposition of the drug in the epithelium of the cornea; approximately two months were required for the drug to disappear from these cells[104]. Patients who had received oral tilorone developed gastrointestinal disorders (diarrhea, vomiting) but no systemic interferon production[104].

The toxic properties of ds–RNAs such as poly(I) · poly(C) are reminiscent of those of bacterial endotoxin (pyrogenicity, and multivalent toxic effects on thymus, spleen, liver, bone marrow, gut, brain, eye, . . . , as reviewed extensively in Refs.[4, 199]). This toxicity is even more pronounced in virus-infected animals[200]. Is it possible to uncouple the interferon-inducing (or antiviral) activities of ds–RNAs from their toxic properties, or do they co-vary? Several procedures have been used in attempts to increase the interferon-inducing capacity of poly(I) · poly(C) in mice: *e.g.* simultaneous injection of lead acetate or cycloheximide, and prior injection of Freund's adjuvant or chlorite-oxidized oxyamylose (COAM)[201]. These procedures brought about a parallel increase in interferon production and toxicity, suggesting that the two effects could not be separated. In other experiments, the poly(I) · poly(C) molecule itself was chosen as target for the modifications, and the modifications carried out were aimed at decreasing the toxicity rather than increasing the activity of the ds–RNA, viz. reduction of the molecular size of poly(C)[111, 121], and introduction of unpaired bases (G or U) in the poly(C) strand[112, 202]. These modifications did not markedly affect the antiviral activity and interferon-inducing capacity of poly(I) · poly(C), yet decreased its toxicity to a marked extent[121, 202]. These findings suggest that the interferon-inducing and toxic properties of poly(I) · poly(C) do not necessarily co-vary but can be uncoupled by appropriate modifications in the molecule. Further studies along these lines may eventually help to resolve one of the oldest questions in interferon research:

is interferon production a specific cellular response or merely the consequence of a toxic alteration?

7. Summary

Various synthetic substances are able to stimulate interferon production, some of them only *in vivo* (polycarboxylates: pyran copolymer, polyacrylic acid, chlorite-oxidized oxyamylose; low molecular weight compounds: tilorone, ...), others *in vivo* and *in vitro* [polynucleotides: poly(I) · poly(C), poly(A) · poly(U) and other ds–RNAs].

The interferon-inducing capacity of polycarboxylates and polynucleotides depends on a number of structural requirements. These requirements have been most extensively studied with synthetic polynucleotides. Keynotes for activity are: high molecular weight, highly ordered double-stranded structure (high thermal stability), adequate resistance to degradation by nucleases, and the presence of 2'-hydroxyl groups. Development of new compounds with high specific activity should adhere to these guidelines. Similar structural requirements (high molecular weight, structural stability, polyanionic character, and, a rather vaguely understood spatial configuration) govern the activity of polycarboxylates. It has so far been impossible to identify the molecular determinants for activity in the low molecular weight interferon inducers. The sole characteristic they have in common, is that they are all diamines; the presence of this requirement does, of course, not allow to predict whether the congener would be active or not.

The mechanism of interferon production has been investigated most thoroughly with ds–RNAs. It has gained general acceptance that the production of interferon is based on a genetic derepression mechanism. Some elements of this genetic regulation system have been located or identified, *e.g.* the structural genes for interferon synthesis and the interferon messenger RNA. Other elements have only been demonstrated indirectly: the repressor protein (identical to the purported receptor molecule, that is the molecule to which the interferon inducer should combine to trigger the interferon response?), and the mode of action of the repressor (inhibition of interferon synthesis at the translational level?).

In vivo, interferon inducers exert a wide variety of effects on host defense mechanisms: interferon production, pyrogenicity, stimulation of cellular and humoral immunity and stimulation of reticulo-endothelial activity. It is not surprising, therefore, that interferon inducers increase the host's resistance to both viral and non-viral (bacterial, fungal, protozoal) infections and tumor growth. A major obstacle confronting the clinical usefulness of interferon inducers is their toxicity. Are interferon induction and toxicity necessarily coupled or can they be disconnected? Preliminary evidence suggests that, at least with poly(I) · poly(C), toxicity and activity could be uncoupled.

E. De Clercq

8. References

1) Isaacs, A., Lindenmann, J.: Proc. Roy. Soc. *B147*, 258 (1957).
2) Finter, N. B. (ed.): Interferons and interferon inducers. Amsterdam: Elsevier/Excerpta Medica/North-Holland Associated Scientific Publishers 1973.
3) De Clercq, E., Merigan, T. C.: Annu. Rev. Med. *21*, 17 (1970).
4) De Clercq, E., Merigan, T. C.: Arch. Internal Med. *126*, 94 (1970).
5) De Clercq, E.: Mechanism of the antiviral activity of synthetic polyanions. Thesis, University of Leuven 1971.
6) Colby, C., Morgan, M. J.: Annu. Rev. Microbiol. *25*, 333 (1971).
7) Colby, C., Jr.: Progr. Nucl. Acid Res. Mol. Biol. *II*, 1 (1971).
8) Kleinschmidt, W. J.: Annu. Rev. Biochem. *41*, 517 (1972).
9) Grossberg, S. E.: New Engl. J. Med. *287*, 13 (1972); *287*, 79 (1972); *287*, 122 (1972).
10) Rodgers, R., Merigan, T. C.: CRC Critical Reviews in Clinical Laboratory Sciences *3*, 131 (1972).
11) De Clercq, E., in: Virus-cell interactions and viral antimetabolites (ed. D. Shugar), p. 65. New York: Academic Press 1972.
12) Carter, W. A. (ed.): Selective inhibitors of viral functions. Cleveland: The Chemical Rubber Company 1973.
13) De Clercq, E., in: Selective inhibitors of viral functions (ed. W. A. Carter). Cleveland: The Chemical Rubber Company 1973.
14) De Clercq, E., Stewart, W. E. II, in: Selective inhibitors of viral functions (ed. W. A. Carter). Cleveland: The Chemical Rubber Company 1973.
15) Carter, W. A.: Proc. Nat. Acad. Sci. *67*, 620 (1970).
16) Green, J. A., Cooperband, S. R., Kibrick, S.: Science *164*, 1415 (1969).
17) Stinebring, W. R., Absher, P. M.: Ann. N. Y. Acad. Sci. *173*, 714 (1970).
18) Salvin, S. B., Youngner, J. S., Lederer, W. H.: Infect. Immun. *7*, 68 (1973).
19) Youngner, J. S.: Infect. Immun. *6*, 646 (1972).
20) Youngner, J. S.: Virology *40*, 335 (1970).
21) Youngner, J. S., Stinebring, W. R., Taube, S. E.: Virology *27*, 541 (1965).
22) De Clercq, E., Merigan, T. C.: Virology *42*, 799 (1970).
23) Falcoff, R., Falcoff, E.: Biochim. Biophys. Acta *182*, 501 (1969).
24) Falcoff, R., Falcoff, E.: Biochim. Biophys. Acta *199*, 147 (1970).
25) Colby, C., Duesberg, P. H.: Nature *222*, 940 (1969).
26) Lockart, R. Z., Jr., Bayliss, N. L., Toy, S. T., Yin, F. H.: J. Virol. *2*, 962 (1968).
27) Lomniczi, B., Burke, D. C.: J. Gen. Virol. *8*, 55 (1970).
28) Goorha, R. M., Gifford, G. E.: Proc. Soc. Exp. Biol. Med. *134*, 1142 (1970).
29) Dianzani, F., Gagnoni, S., Buckler, C. E., Baron, S.: Proc. Soc. Exp. Biol. Med. *133*, 324 (1970).
30) Gandhi, S. S., Burke, D. C.: J. Gen. Virol. *6*, 95 (1970).
31) Gandhi, S. S., Burke, D. C., Scholtissek, C.: J. Gen. Virol. *9*, 97 (1970).
32) Bakay, M., Burke, D. C.: J. Gen. Virol. *16*, 399 (1972).
33) Clavell, L. A., Bratt, M. A.: J. Virol. *8*, 500 (1971).
34) Meager, A., Burke, D. C.: Nature *235*, 280 (1972).
35) Montagnier, L. M.: C. R. Acad. Sci. (Paris) *267*, 1417 (1968).
36) Harel, L., Montagnier, L.: Nature New Biol. *229*, 106 (1971).
37) De Maeyer, E., De Maeyer-Guignard, J., Montagnier, L.: Nature New Biol. *229*, 109 (1971).
38) Kimball, P. C., Duesberg, P. H.: J. Virol. *7*, 697 (1971).
39) Epstein, L. B., Cline, M. J., Merigan, T. C.: J. Clin. Invest. *50*, 744 (1971).
40) Epstein, L. B., Stevens, D. A., Merigan, T. C.: Proc. Nat. Acad. Sci. *69*, 2632 (1972).
41) Haber, J., Rosenau, W., Goldberg, M.: Nature New Biol. *238*, 60 (1972).
42) Cooper, H. L., Rubin, A. D.: Science *152*, 516 (1966).
43) Torelli, U. L., Henry, P. H., Weissman, S. M.: J. Clin. Invest. *47*, 1083 (1968).
44) Kay, J. E., Cooper, H. L.: Biochim. Biophys. Acta *186*, 62 (1969).

[45] Regelson, W. E., in: The reticulo-endothelial system and atherosclerosis (eds. N. R. Di Luzio and R. Paoletti), p. 315. New York: Plenum Press 1967.

[46] Merigan, T. C.: Nature *214*, 416 (1967).

[47] Merigan, T. C., Regelson, W.: New Engl. J. Med. *277*, 1283 (1967).

[48] Merigan, T. C., Finkelstein, M. S.: Virology *35*, 363 (1968).

[49] De Somer, P., De Clercq, E., Billiau, A., Schonne, E., Claesen, M.: J. Virol. *2*, 886 (1968).

[50] Niblack, J. F.: Ann. N. Y. Acad. Sci. *173*, 536 (1970).

[51] Claes, P., Billiau, A., De Clercq, E., Desmyter, J., Schonne, E., Vanderhaeghe, H., De Somer, P.: J. Virol. *5*, 313 (1970).

[52] Came, P. E., Lieberman, M., Pascale, A., Shimonaski, G.: Proc. Soc. Exp. Biol. Med. *131*, 443 (1969).

[53] De Clercq, E., Eckstein, F., Merigan, T. C.: Ann. N. Y. Acad. Sci. *173*, 444 (1970).

[54] Suzuki, S., Suzuki, M., Chaki, F.: Jap. J. Microbiol. *16*, 1 (1972).

[55] Field, A. K., Tytell, A. A., Lampson, G. P., Hilleman, M. R.: Proc. Nat. Acad. Sci. *58*, 1004 (1967).

[56] Field, A. K., Tytell, A. A., Lampson, G. P., Hilleman, M. R.: Proc. Nat. Acad. Sci. *61*, 340 (1968).

[57] Vilcek, J., Ng, M. H., Friedman-Kien, A. E., Krawciw, T.: J. Virol. *2*, 648 (1968).

[58] Finkelstein, M. S., Bausek, G. H., Merigan, T. C.: Science *161*, 465 (1968).

[59] Youngner, J. S., Hallum, J. V.: Virology *35*, 177 (1968).

[60] Falcoff, E., Perez-Bercoff, R.: Biochim. Biophys. Acta *174*, 108 (1969).

[61] De Clercq, E., Merigan, T. C.: Nature *222*, 1148 (1969).

[62] Colby, C., Chamberlin, M. J.: Proc. Nat. Acad. Sci. *63*, 160 (1969).

[63] Bausek, G. H., Merigan, T. C.: Virology *39*, 491 (1969).

[64] De Clercq, E., Nuwer, M. R., Merigan, T. C.: J. Clin. Invest. *49*, 1565 (1970).

[65] Du Buy, H. G., Johnson, M. L., Buckler, C. E., Baron, S.: Proc. Soc. Exp. Biol. Med. *135*, 340 (1970).

[66] Buckler, C. E., Du Buy, H. G., Johnson, M. L., Baron, S.: Proc. Soc. Exp. Biol. Med. *136*, 394 (1971).

[67] Field, A. K., Young, C. W., Krakoff, I. H., Tytell, A. A., Lampson, G. P., Nemes, M. M., Hilleman, M. R.: Proc. Soc. Exp. Biol. Med. *136*, 1180 (1971).

[68] De Clercq, E.: Proc. Soc. Exp. Biol. Med. *141*, 340 (1972).

[69] De Clercq, E., Eckstein, F., Merigan, T. C.: Science *165*, 1137 (1969).

[70] De Clercq, E., Wells, R. D., Merigan, T. C.: Nature *226*, 364 (1970).

[71] De Clercq, E., Eckstein, F., Sternbach, H., Merigan, T. C.: Virology *42*, 421 (1970).

[72] De Clercq, E., Wells, R. D., Grant, R. C., Merigan, T. C.: J. Mol. Biol. *56*, 83 (1971).

[73] De Clercq, E., Merigan, T. C.: J. Gen. Virol. *10*, 125 (1971).

[74] De Clercq, E., Wells, R. D., Merigan, T. C.: Virology *47*, 405 (1972).

[75] Black, D. R., Eckstein, F., De Clercq, E., Merigan, T. C.: Antimicrob. Ag. Chemother. *3*, 198 (1973).

[76] Baron, S., Bogomolova, N. N., Billiau, A., Levy, H. B., Buckler, C. E., Stern, R., Naylor, R.: Proc. Nat. Acad. Sci. *64*, 67 (1969).

[77] Billiau, A., Buckler, C. E., Dianzani, F., Uhlendorf, C., Baron, S.: Proc. Soc. Exp. Biol. Med. *132*, 790 (1969).

[78] Matsuda, S., Kida, M., Shirafuji, H., Yoneda, M., Yaoi, H.: Arch. ges. Virusforsch. *34*, 105 (1971).

[79] Black, D. R., Eckstein, F., Hobbs, J. B., Sternbach, H., Merigan, T. C.: Virology *48*, 537 (1972).

[80] Pitha, J., Pitha, P. M.: Science *172*, 1146 (1971).

[81] De Clercq, E., Janik, B.: Biochim. Biophys. Acta *324*, 50 (1973).

[82] Torrence, P. F., Waters, J. A., Buckler, C. E., Witkop, B.: Biochem. Biophys. Res. Commun. *52*, 890 (1973).

[83] Steward, D. L., Herndon, W. C., Jr., Schell, K. R.: Biochim. Biophys. Acta *262*, 227 (1972).

[84] De Clercq, E., Zmudzka, B., Shugar, D.: FEBS Letters *24*, 137 (1972).

85) Krueger, R. F., Mayer, G. D.: Science *169*, 1213 (1970).
86) Mayer, G. D., Krueger, R. F.: Science *169*, 1214 (1970).
87) De Clercq, E., Merigan, T. C.: J. Infect. Dis. *123*, 190 (1971).
88) Camyre, K. P., Groelke, J. W., Mayer, G. D., Krueger, R. F.: Personal communication, Interferon Scientific Memorandum n °550 (1971).
89) Stringfellow, D. A., Glasgow, L. A.: Antimicrob. Ag. Chemother. *2*, 73 (1972).
90) Hofmann, H., Kunz, C.: Arch. ges. Virusforsch. *37*, 262 (1972).
91) Krueger, R. F., Mayer, G. D., Camyre, K. P., Yoshimura, S.: Submitted for publication (1973).
92) Diederich, J., Lodemann, E., Wacker, A.: Naturwissenschaften *59*, 172 (1972).
93) Diederich, J., Lodemann, E., Wacker, A.: Arch. ges. Virusforsch. *40*, 82 (1973).
94) Glaz, E. T., Szolgay, E., Stöger, I., Talas, M.: Antimicrol. Ag. Chemother. *3*, 537 (1973).
95) Hoffman, W. W., Korst, J. J., Niblack, J. F., Cronin, T. H.: Antimicrol. Ag. Chemother. *3*, 498 (1973).
96) Lukas, B., Hruskova, J.: Acta Virol. *12*, 263 (1968).
97) De Somer, P., De Clercq, E., Billiau, A., Schonne, E., Claesen, M.: J. Virol. *2*, 878 (1968).
98) Dennis, A. J., Wilson, H. E., Barker, A. D., Rheins, M. S.: Proc. Soc. Exp. Biol. Med. *141*, 782 (1972).
99) Fantes, K. H., Taylor, M. M., Burman, C. J.: Personal communication, Interferon Scientific Memorandum n °658 (1972).
100) Vandeputte, M., De Clercq, E., Billiau, A., Claesen, M., De Somer, P., in: US-Japan Cooperative Seminar on Interferon (eds. Y. Nagano and H. B. Levy), p. 206. Tokyo: Igaku Shoin Ltd. 1970.
101) De Clercq, E., De Somer, P.: Infect. Immun., in press (1973).
102) Zinober, M. R., Pirtle, E. C.: Personal communication, Interferon Scientific Memorandum n° 509 (1971).
103) Portnoy, J., Merigan, T. C.: J. Infect. Dis. *124*, 545 (1971).
104) Kaufman, H. E., Centifanto, Y. M., Ellison, E. D., Brown, D. C.: Proc. Soc. Exp. Biol. Med. *137*, 357 (1971).
105) De Clercq, E., De Somer, P.: Appl. Microbiol. *16*, 1314 (1968).
106) Mohr, S. J., Brown, D. G., Coffey, D. S.: Nature New Biol. *240*, 250 (1972).
107) Remington, J. S., Merigan, T. C.: Nature *226*, 361 (1970).
108) Billiau, A., Desmyter, J., De Somer, P.: J. Virol. *5*, 321 (1970).
109) Lampson, G. P., Field, A. K., Tytell, A. A., Nemes, M. M., Hilleman, M. R.: Proc. Soc. Exp. Biol. Med. *135*, 911 (1970).
110) Shiokawa, K., Yaoi, H.: Arch. ges. Virusforsch. *38*, 109 (1972).
111) Tytell, A. A., Lampson, G. P., Field, A. K., Nemes, M. M., Hilleman, M. R.: Proc. Soc. Exp. Biol. Med. *135*, 917 (1970).
112) Carter, W. A., Pitha, P. M., Marshall, L. W., Tazawa, I., Tazawa, S., Ts'O, P. O. P.: J. Mol. Biol. *70*, 567 (1972).
113) Wacker, A., Singh, A., Svec, J., Lodemann, E.: Naturwissenschaften *56*, 638 (1969).
114) Niblack, J. F., McCreary, M. B.: Nature *233*, 52 (1971).
115) Morahan, P. S., Munson, A. E., Regelson, W., Commerford, S. L., Hamilton, L. D.: Proc. Nat. Acad. Sci. *69*, 842 (1972).
116) Doty, P., Boedtker, H., Fresco, J. R., Haselkorn, R., Litt, M.: Proc. Nat. Acad. Sci. *45*, 482 (1959).
117) Gresser, I.: Personal communication, Interferon Scientific Memorandum n °143 (1969).
118) Harnden, M. R., Brown, A. G., Vere Hodge, R. A., Planterose, D. N.: Personal communication, Interferon Scientific Memorandum n °715 (1972).
119) Pitha, P. M., Carter, W. A.: Nature New Biol. *234*, 105 (1971).
120) Wang, A. C., Kallenbach, N. R.: J. Mol. Biol. *62*, 591 (1971).
121) Lampson, G. P., Nemes, M. M., Field, A. K., Tytell, A. A., Hilleman, M. R.: Proc. Soc. Exp. Biol. Med. *141*, 1068 (1972).

122) De Clercq, E., De Somer, P.: J. Virol. *9*, 721 (1972).
123) De Clercq, E., De Somer, P.: Science *173*, 260 (1971).
124) De Clercq, E., Stewart, W. E. II, De Somer, P.: Virology *54*, 278 (1973).
125) Dianzani, F., Cantagalli, P., Gagnoni, S., Rita, G.: Proc. Soc. Exp. Biol. Med. *128*, 708 (1968).
126) Lampson, G. P., Tytell, A. A., Field, A. K.; Nemes, M. M., Hilleman, M. R.: Proc. Soc. Exp. Biol. Med. *132*, 212 (1969).
127) Dianzani, F., Gagnoni, S., Cantagalli, P.: Ann. N. Y. Acad. Sci. *173*, 727 (1970).
128) Tilles, J. G.: Proc. Soc. Exp. Biol. Med. *133*, 1334 (1970).
129) Mecs, I., Rosztoczy, I.: Acta Virol. *15*, 280 (1971).
130) Pitha, P., Carter, W. A.: Virology *45*, 777 (1971).
131) Dianzani, F., Baron, S., Buckler, C. E., Levy, H. B.: Proc. Soc. Exp. Biol. Med. *136*, 1111 (1971).
132) Vilcek, J., Barmak, S. L., Havell, E. A.: J. Virol. *10*, 614 (1972).
133) Pitha, P. M., Pitha, J.: J. Gen. Virol., in press (1973).
134) Dianzani, F., Rita, G., Cantagalli, P., Gagnoni, S.: J. Immunol. *102*, 24 (1969).
135) Finter, N. B., in: L'Interféron, Colloques de l'Institut National de la Santé et de la Recherche Médicale, n °6., p. 325. INSERM (Paris) 1970.
136) Rice, J. M., Turner, W., Chirigos, M. A., Rice, N. R.: Appl. Microbiol. *19*, 867 (1970).
137) Rice, J. M., Turner, W., Chirigos, M. A., Spahn, G.: Appl. Microbiol. *22*, 380 (1971).
138) Vilcek, J., Rossman, T. G., Varacalli, F.: Nature *222*, 682 (1969).
139) Vilcek, J.: Ann. N. Y. Acad. Sci. *173*, 390 (1970).
140) Vilcek, J., Ng, M. H.: J. Virol. *7*, 588 (1971).
141) Tan, Y. H., Armstrong, J. A., Ke, Y. H., Ho, M.: Proc. Nat. Acad. Sci. *67*, 464 (1970).
142) Tan, Y. H., Armstrong, J. A., Ho, M.: Virology *3*, 503 (1971).
143) Ho, M., Tan, Y. H., Armstrong, J. A.: Proc. Soc. Exp. Biol. Med. *139*, 259 (1972).
144) Myers, M. W., Friedman, R. M.: J. Nat. Cancer Inst. *47*, 757 (1971).
145) Havell, E. A., Vilcek, J.: Antimicrob. Ag. Chemother. *2*, 476 (1972).
146) Billiau, A., Joniau, M., De Somer, P.: J. Gen. Virol. *19*, 1 (1973).
147) Rosztoczy, I., Mecs, I.: Acta Virol. *14*, 398 (1970).
148) Stewart, W. E. II, Gosser, L. B., Lockart, R. Z., Jr.: J. Virol. *7*, 792 (1971).
149) Stewart, W. E. II, Gosser, L. B., Lockart, R. Z., Jr.: J. Gen. Virol. *13*, 35 (1971).
150) Stewart, W. E. II, Gosser, L. B., Lockart, R. Z., Jr.: J. Gen. Virol. *15*, 85 (1972).
151) Stewart, W. E. II, De Clercq, E., Billiau, A., Desmyter, J., De Somer, P.: Proc. Nat. Acad. Sci. *69*, 1851 (1972).
152) Margolis, S. A., Oie, H., Levy, H. B.: J. Gen. Virol. *15*, 119 (1972).
153) De Clercq, E., Stewart, W. E. II, De Somer, P.: Infect. Immun. *8*, 309 (1973).
154) Billiau, A., Van den Berghe, H., De Somer, P.: J. Gen. Virol. *14*, 25 (1972).
155) Nordlund, J. J., Wolff, S. M., Levy, H. B.: Proc. Soc. Exp. Biol. Med. *133*, 439 (1970).
156) Stern, R.: Biochem. Biophys. Res. Commun. *41*, 608 (1970).
157) Friedman, R. M., Barth, R. F., Stern, R.: Nature New Biol. *230*, 17 (1971).
158) Janik, B., Kotick, M. P., Kreiser, T. H., Reverman, L. F., Sommer, R. G., Wilson, D. P.: Biochem. Biophys. Res. Commun. *46*, 1153 (1972).
159) Janion, C., Zmudzka, B., Shugar, D.: Acta Biochim. Polon. *17*, 31 (1970).
160) Zmudzka, B., Shugar, D.: Acta Biochim. Polon. *18*, 321 (1971).
161) Dunlap, B. E., Friderici, K. H., Rottman, F.: Biochemistry *10*, 2581 (1971).
162) Hobbs, J., Sternbach, H., Sprinzl, M., Eckstein, F.: Biochemistry *11*, 4336 (1972).
163) Torrence, P. F., Waters, J. A., Witkop, B.: J. Am. Chem. Soc. *94*, 3638 (1972).
164) Hobbs, J., Sternbach, H., Eckstein, F.: Biochem. Biophys. Res. Commun. *46*, 1509 (1972).
165) Arnott, S.: Progr. Biophys. Mol. Biol. *21*, 265 (1970).
166) Milman, G., Langridge, R., Chamberlin, M. J.: Proc. Nat. Acad. Sci. *57*, 1804 (1967).
167) Colby, C., Stollar, B. D., Simon, M. I.: Nature New Biol. *229*, 172 (1971).
168) Arnott, S., Fuller, W., Hodgson, A., Prutton, I.: Nature *220*, 561 (1968).
169) Arnott, S., Hukins, D. W. L.: Nature *224*, 886 (1969).

170) Chandra, P., Zunino, F., Zaccara, A., Wacker, A., Götz, A.: FEBS Letters *23*, 145 (1972).
171) Chandra, P., Zunino, F., Gaur, V. P., Zaccara, A., Woltersdorf, M., Luoni, G., Götz, A.: FEBS Letters *28*, 5 (1972).
172) Field, A. K., Lampson, G. P., Tytell, A. A., Nemes, M. M., Hilleman, M. R.: Proc. Nat. Acad. Sci. *58*, 2102 (1967).
173) Jacob, F., Monod, J.: J. Mol. Biol. *3*, 318 (1961).
174) Kleinschmidt, W. J., Cline, J. C., Murphy, E. B.: Proc. Nat. Acad. Sci. *52*, 1751 (1964).
175) Ellis, L. F., Kleinschmidt, W. J.: Nature *215*, 649 (1967).
176) Kleinschmidt, W. J., Ellis, L. F., Van Frank, R. M., Murphy, E. B.: Nature *220*, 167 (1968).
177) Cassingena, R., Chany, C., Vignal, M., Suarez, H., Lazar, P.: Proc. Nat. Acad. Sci. *68*, 580 (1971).
178) De Maeyer-Guignard, J., De Maeyer, E., Montagnier, L.: Proc. Nat. Acad. Sci. *69*, 1203 (1972).
179) De Clercq, E., De Somer, P.: J. Gen. Virol. *19*, 113 (1973).
180) Field, A. K., Tytell, A. A., Lampson, G. P., Hilleman, M. R.: Proc. Soc. Exp. Biol. Med. *140*, 710 (1972).
181) De Clercq, E., De Somer, P.: J. Gen. Virol. *16*, 435 (1972).
182) De Clercq, E., De Somer, P.: Submitted for publication (1973).
183) Wagner, A. F., Bugianesi, R. L., Shen, T. Y.: Biochem. Biophys. Res. Commun. *45*, 184 (1971).
184) Pitha, P. M., Pitha, J.: J. Gen. Virol., in press (1973).
185) Stewart, W. E. II, Scott, W., Sulkin, S. E.: J. Virol. *4*, 147 (1969).
186) Schafer, T. W., Lockart, R. Z., Jr.: Nature *226*, 449 (1970).
187) Vilcek, J., Varacalli, F.: J. Gen. Virol. *13*, 185 (1971).
188) Kjeldsberg, E., Flikke, M.: J. Gen. Virol. *10*, 147 (1971).
189) Burke, D. C., in: Interferons (ed. N. B. Finter), p. 55. Amsterdam: North-Holland Publishing Co. 1966.
190) Tomkins, G. M., Gelehrter, T. D., Granner, D., Martin, D. Jr., Samuels, H. H., Thompson, E. B.: Science *166*, 1474 (1969).
191) Nahon, E., Michelson, A. M., Lacour, F.: Biochim. Biophys. Acta *149*, 127 (1967).
192) Schwartz, E. F., Stollar, B. D.: Biochem. Biophys. Res. Commun. *35*, 115 (1969).
193) Schur, P. H., Monroe, M.: Proc. Nat. Acad. Sci. *63*, 1108 (1969).
194) Dianzani, F., Forni, G., Ponzi, A. N., Pugliese, A., Cavallo, G.: Proc. Soc. Exp. Biol. Med. *139*, 93 (1972).
195) Field, A. K., Tytell, A. A., Lampson, G. P., Hilleman, M. R.: Proc. Soc. Exp. Biol. Med. *139*, 1113 (1972).
196) Steinberg, A. D., Baron, S., Uhlendorf, C., Talal, N.: Proc. Soc. Exp. Biol. Med. *137*, 558 (1971).
197) Steinberg, A. D., Baron, S., Talal, N.: Proc. Nat. Acad. Sci. *63*, 1102 (1969).
198) Rohovsky, M. W., Newberne, J. W., Gibson, J. P.: J. Toxicol. Appl. Pharmacol. *17*, 556 (1970).
199) De Clercq, E., Stewart, W. E. II: Medikon *1*, 331 (1972).
200) De Clercq, E., Stewart, W. E. II, De Somer, P.: Infect. Immun. *7*, 167 (1973).
201) De Clercq, E., Stewart, W. E. II, De Somer, P.: Infect. Immun. *6*, 344 (1972).
202) Carter, W. A., Marshall, L. W., Ts'O, P. O. P.: Personal communication (1973).

Received May 21, 1973

Synthesis and Properties
of Some New NAD⊕ Analogues

Professor Dr. Dr. Christoph Woenckhaus

Klinikum der Johann Wolfgang Goethe-Universität, Gustav-Embden-Zentrum der Biologischen Chemie, Abteilung für Enzymologie, Frankfurt (Main)

Contents

1. Introduction

The name nicotinamide-adenine dinucleotide (NAD$^\oplus$) describes a structure in which nucleotides containing nicotinamide and adenine are linked through a pyrophosphate bond. The substance serves as coenzyme for a number of dehydrogenases active in the oxidation or reduction of metabolites. The reaction, as written, involves the stoichiometric, reversible transfer of hydrogen from the metabolite to the coenzyme[1, 3]. It is the pyridine ring of NAD$^\oplus$ that changes during oxidation or reduction. The combination of nicotinamide with ribose creates a quaternary nitrogen with a positive charge in the ring, and the nucleotide is a stronger oxidizing agent than the free nicotinamide alone. The additional structure of the molecule beyond the reactive pyridine ring serves to bind it in a distinctive way to specific enzymes. The reduction of NAD$^\oplus$ requires the addition of a hydride ion, a proton with two electrons, and the nucleotide loses its positive charge in the process. The hydride ion is contributed by the substrate that is being oxidized. In addition to the enzymatic reaction, NAD$^\oplus$ can be reduced chemically with sodium hyposulfite. The reduced NAD is not able to transfer its hydrogen in the absence of an enzyme. The coenzyme is activated by a specific interaction with the rest of the enzyme protein. The NAD$^\oplus$ (or NADH) is bound to specific sites on the enzyme molecule and forms reactive coenzyme—enzyme complexes.

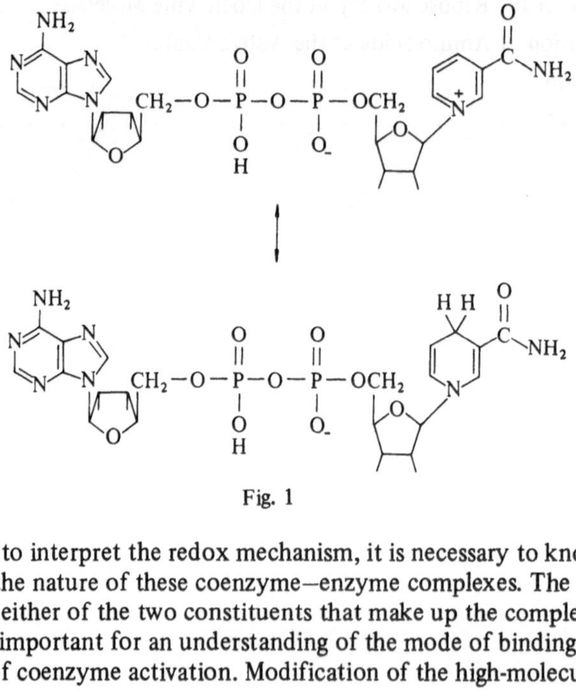

Fig. 1

In order to interpret the redox mechanism, it is necessary to know something about the nature of these coenzyme—enzyme complexes. The chemical alteration of either of the two constituents that make up the complex could therefore be important for an understanding of the mode of binding and the mechanism of coenzyme activation. Modification of the high-molecular weight protein moiety of the enzyme is rather difficult and often provokes total loss of the catalytic activity of the complex; there are very few cases where its activity is conserved[2]. Modification of the coenzyme molecule, however, is readily effected. We have therefore attempted to make structural changes in the nicotinamide-adenine dinucleotide moiety and then to study its interaction with dehydrogenases. These studies have been helpful in revealing the nature of enzyme-coenzyme interactions.

The removal of the adenine ring from the coenzyme molecule gives nicotinamide-riboside-5'-pyrophosphate-5"-ribose, a compound that has little coenzyme activity. The nicotinamide mononucleotide acts as hydrogen acceptor with alcohol dehydrogenase from liver. The reduced forms of both coenzyme fragments react with dehydrogenases and their appropriate substrates. However, compared to that with NADH, the reaction of dehydrogenases with the coenzyme fragment which contains no adenine show that the adenine part of the molecule is important for the enzymatic reaction. If the nonfunctional part of the molecule is necessary for the binding of coenzymes to the active center, its absence might be expected to lead to a higher dissociation of the complexes and hence a reduction of their catalytic activity. Besides being functional for binding, the adenine part of the molecule may also be involved in the activation of nicotinamide when the coenzyme–enzyme complex is formed. Adenine may participate in the transfer of energy from the protein to the functional part of the molecule; this assumption is supported by the fact that, in NAD$^{\oplus}$ and NADH, an interaction between the two rings is observed. The extinction of NAD$^{\oplus}$ is not an additive function of the extinction of adenosine monophosphate and nicotinamide mononucleotide. Only after the splitting of the pyrophosphate bridge is there the rise in extinction that one would expect. The same occurs with NADH: the absorption maxima at 260 nm and 340 nm increase only after dissociation of the pyrophosphate bridge[4]. Moreover, a blue shift of 5 nm is observed when the dihydronicotinamide part is detached. The direct transfer of energy from adenine to the dihydronicotinamide ring of the coenzyme has been reported by Weber[5]. The dihydronicotinamide ring of NADH absorbs light in the range 310 to 390 nm and emits energy over the broad range 390 to 500 nm; the maximum lies at 456 nm (Figs. 2, 3 and 4).

The absorption maximum of adenine occurs at 260 nm. It does not fluoresce at neutral pH at room temperature. On irradiation at 260 nm the dihydronicotinamide part of the molecule emits energy at 456 nm. This process involves the transfer of the energy of light from the adenine ring to the dihydronicotinamide ring, which then emits fluorescent light. This leads us to conclude that both coenzyme forms, NAD$^{\oplus}$ and NADH, lie in a folded conformation and that their rings are arranged in parallel[6]. If the coenzyme were to bind at the active center of the enzyme, there would be an interaction between the coenzyme moieties and the side chains of the enzyme molecule which would then influence the optical properties of the enzyme and the coenzyme.

211

2. Coenzyme—Enzyme Complexes

The optical changes that occur after complex formation between the enzyme and the coenzyme are expressed in the fluorescence spectrum. The excitation spectrum is usually reflected in the absorption spectrum, since the energy taken up in the absorption region is generally emitted at another wavelength. To obtain the excitation spectrum, the irradiation wavelength was varied and the intensity of the light emitted at a constant wavelength was measured. To obtain the emission spectrum, the substance was irradiated at a constant wavelength and the intensity of the light emitted was measured.

Dehydrogenases exhibit a fluorescence similar to that of tryptophan. Irradiation of dehydrogenases at 280 nm gives an emission maximum at 340 nm (Figs. 2 and 3). This pattern is characteristic of enzymes which contain tryptophan and can be designated protein fluorescence. The other side residues, tyrosine or phenylalanine, fluoresce very weakly and probably do not contribute significantly to the fluorescence of the enzyme[7].

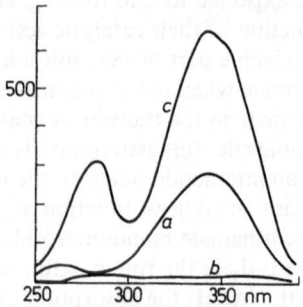

Fig. 2. Fluorescence excitation spectrum of (a) NADH, $c = 1 \times 10^{-4}$M, (b) alcohol dehydrogenase from yeast, $c = 5 \times 10^{-5}$ M, (c) binary complex of NADH and alcohol dehydrogenase. Emission 455 nm

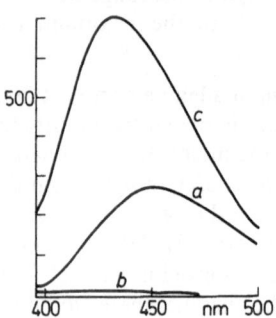

Fig. 3. Fluorescence emission spectrum of (a) NADH, $c = 1 \times 10^{-4}$M, (b) alcohol dehydrogenase from yeast, $c = 5 \times 10^{-5}$M, (c) binary complex. Excitation: 340 nm

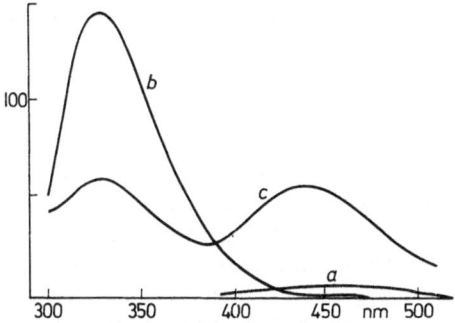

Fig. 4. Fluorescence emission spectrum of (a) NADH, $c = 1 \times 10^{-4}$ M, (b) alcohol dehydrogenase from yeast, $c = 5 \times 10^{-5}$ M, (c) binary complex. Excitation: 280 nm, pH 7.5.
Ordinate: wavelength
Abscissa: Intensity of the fluorescent light in arbitrary units

NAD$^\oplus$ does not show fluorescence, whereas in NAD$^\oplus$ or NADH dehydrogenase complexes the fluorescence at 340 nm is reduced after irradiation at 280 nm. In the excitation spectrum of NADH—enzyme complexes increased fluorescence of the coenzyme is recorded after irradiation at 340 nm (Figs. 2, 3 and 4). The emission maximum is shifted by 30 nm from 456 to 426 nm[8]. In the excitation spectrum a new fluorescence band is seen at 280 nm; this band cannot be attributed either to the protein or to the coenzyme, since their emission maximum lies at 426 nm and represents the coenzyme fluorescence after formation of the complex. The coenzyme has very low absorption at 280 nm whereas the tryptophan residue absorbs very strongly at this wavelength, or in this region. However, the emission maximum for tryptophan lies at 340 nm[9]: the irradiated energy is accepted by the tryptophan residue and transferred to the dihydronicotinamide ring, which then emits it[8]. This band is called the energy-transfer band. The depression of protein fluorescence can be explained by the fact that almost all the energy absorbed by the protein at 280 nm is emitted by NADH. The increase in the coenzyme fluorescence is perhaps due to a rigid binding of the coenzyme at the active center[9]. This strong binding may lead to repression of the process causing inactivation of the excited state and reduction of the quantum yield[9, 10]. The blue shift of the coenzyme fluorescence after its binding to the enzyme could be ascribed to a small polarized region in which the dihydronicotinamide ring is present at the active center of the enzyme[7].

The NAD$^\oplus$—enzyme complex exhibits absorption regions that characterize protein, adenine and nicotinamide, i.e. in the range 240 to 290 nm. In the NADH—enzyme complex, the dihydronicotinamide part absorbs at 340 nm. In difference spectra comparing NADH—dehydrogenase complexes and isolated components, the absorption changes between 340 and 360 nm can be attributed to hydrogen transfer[10]. Using coenzyme analogues in which the nonfunctional part shows a different spectral behavior, one observes the changes

213

Ch. Woenckhaus

in absorption after complex formation. These changes can be correlated with single components of the analogues.

3. Coenzyme Analogues

The substitution of the nicotinamide part of the NAD$^\oplus$ molecule by other pyridine rings through an enzyme-catalyzed exchange reaction with NAD-glycohydrolase has been reported[11]. Using this reaction, Kaplan was able to synthesize a number of structural analogues of the coenzyme[12]. Since the functional part of these analogues is modified, they have different redox potentials. Because no direct exchange of charge takes place between dihydro-pyridine and pyridine, it is difficult to measure the redox potential. However,

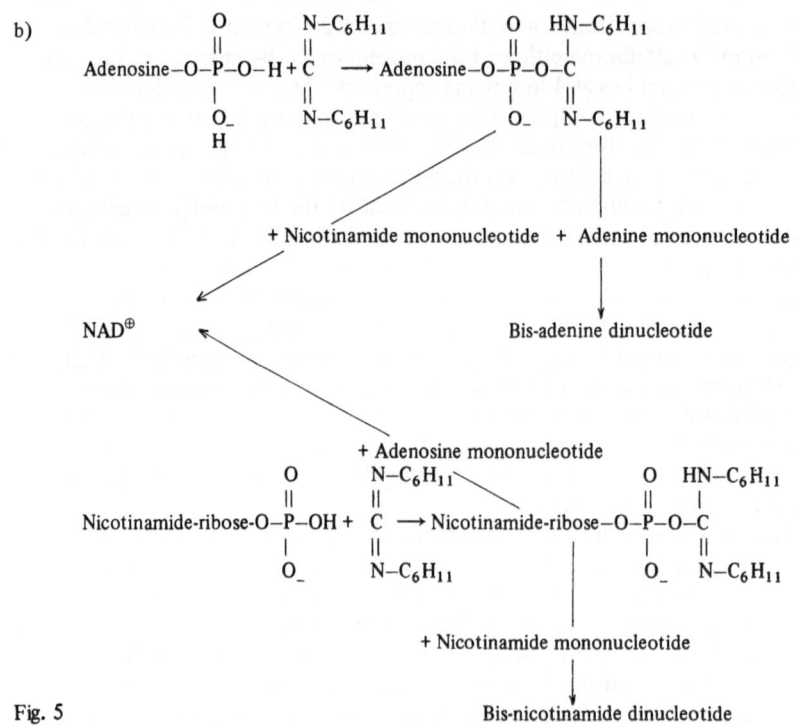

Fig. 5

214

the coenzyme analogues form cyanide adducts with variable dissociation constants and from these dissociation constants one can calculate the approximate redox potentials[13]. Modifications of the nonfunctional part, though they do not alter the redox potential, require the synthesis of the nonfunctional part of the coenzyme which can then be attached by bonding the nucleotide-5'-phosphate with nicotinamide mononucleotide through an anhydride bond[14]. The synthesis of these compounds was effected as follows:

a) Activation of the nucleotide monophosphate. *e.g.* formation of phosphomorpholidate and condensation with nicotinamide mononucleotide[15].
b) Condensation of both nucleotide monophosphates via N,N-dicyclohexyl-carbodiimide in aqueous pyridine[16].

Procedure a) gives a good yield, but it is usually difficult to separate the coenzyme analogues from the starting materials. The formation of symmetrical byproducts is not observed. With procedure b), yields of coenzyme analogues seldom exceed 35%; the symmetrical nucleotide anhydrides formed can easily be separated by ion-exchange chromatography (Fig. 5).

Nucleotide monophosphate can be prepared from naturally occurring nucleotides by substitution reactions, as described earlier[17]. An other possibility is to synthesize the nonfunctional ring, convert it into its heavy-metal salt, and condense with 1-chloro-2,3,5-0-triacetyl ribofuranose[18]. The preparation of nucleosides by the fusion method requires the use of 1,2,3,5-0-tetraacetyl ribofuranose, which in the presence of a catalyst converts the base into a nucleoside[19]. After hydrolysis of the protective acetyl groups, the 2',3' hydroxyl groups of the sugar can be selectively converted with acetone into the isopropylidene nucleoside. The phosphorylation of free 5'-hydroxyl groups with cyanoethyl phosphate and N,N-dicyclohexylcarbodiimide leads to the formation of nucleoside-5' phosphate[20]. In a weakly acidic medium the isopropylidene residue is again split. The nonprotected nucleosides can be selectively phosphorylated at the 5' position with phosphorus oxychloride in a solution of trimethyl phosphate[21] (Fig. 6). The role of the nonfunctional

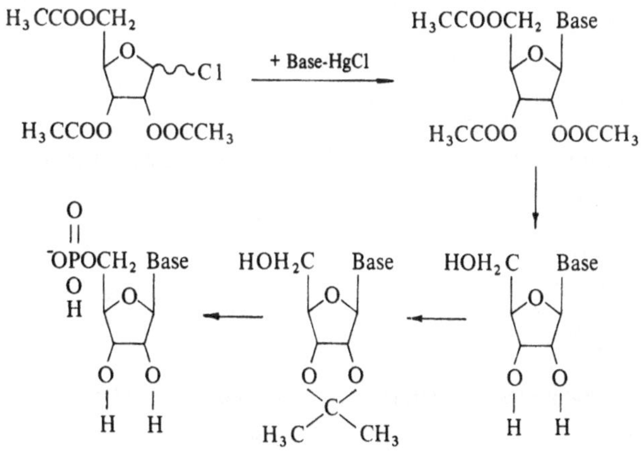

Fig. 6

ring and the ribosyl residue of the coenzyme in binding and activating the nicotinamide part was investigated with these coenzyme analogues. By inserting reactive groups in the coenzyme analogues one can prepare inactivators which, because of their structural similarity to the natural coenzyme, can be attached to the active center and react with side chains of the protein to form covalent bonds. If now one degrades the inactivated enzyme protein, it is possible to identify the amino acids involved in binding the coenzyme.

The removal of adenine from $NAD^{\oplus}$ induces strong inhibition of the catalytic activity of the coenzyme fragment[26]; substitution of the adenine by other purine derivatives gives rise to coenzyme analogues with coenzyme properties almost similar to those of $NAD^{\oplus}$ and NADH[22, 23] (Table 1).

Table 1. Coenzyme properties of analogues containing different purine systems

Enzymes	L-ADH		Y-ADH		LDH	
	$K_M \times 10^4$	V_{max}	$K_M \times 10^4$	V_{max}	$K_M \times 10^4$	V_{max}
Nicotinamide-6-methyl-purine dinucleotide	0.3	500	3.5	20000	1.5	18000
Nicotinamide-purine di-nucleotide	1.2	300	2.5	20000	1.8	18000
Nicotinamide-6-thiopurine dinucleotide	1.6	230	6	10000	3	14000
Nicotinamide-6-methyl-,thiopurine dinucleotide	2.3	300	6.7	10000	2	10000
Nicotinamide-6-methyl-2-chloropurine dinu-cleotide	0.15	330	3	20000	1.5	20000
$NAD^{\oplus}$	0.3	410	2.5	55000	0.75	17000

Abbreviations: L-ADH — alcohol dehydrogenase of liver;
Y-ADH — alcohol dehydrogenase from yeast;
LDH — lactate dehydrogenase;
MDH_c — cycloplasmatic malate dehydrogenase;
MDH_m — mitochondrial malate dehydrogenase;
GAPDH — glyceraldehyde-3-phosphate dehydrogenase.

Some of these derivatives, however, differ from the natural coenzyme in their intramolecular interaction. The hypochromic effect is reduced as compared to $NAD^{\oplus}$ or NADH[24]. The fluorescence spectra of some excited molecules do not show the energy-transfer bands observed with the natural coenzyme[23] (Fig. 7).

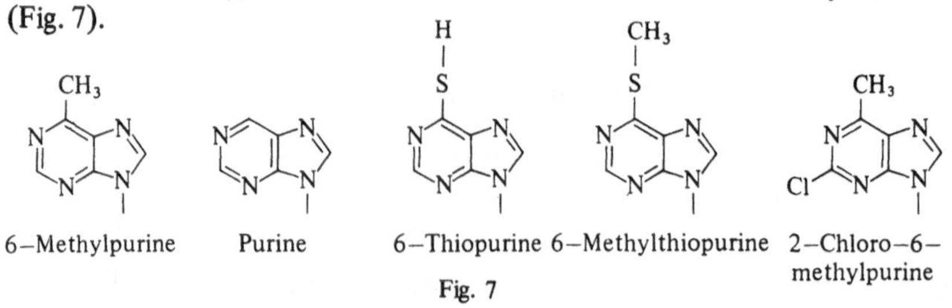

6—Methylpurine   Purine   6—Thiopurine 6—Methylthiopurine 2—Chloro—6—methylpurine

Fig. 7

It has been shown that substitution of the purine ring by other ring systems strongly influences the catalytic activity of the analogues[24]. We have succeeded in substituting the adenine ring by the ring systems shown in Fig. 8.

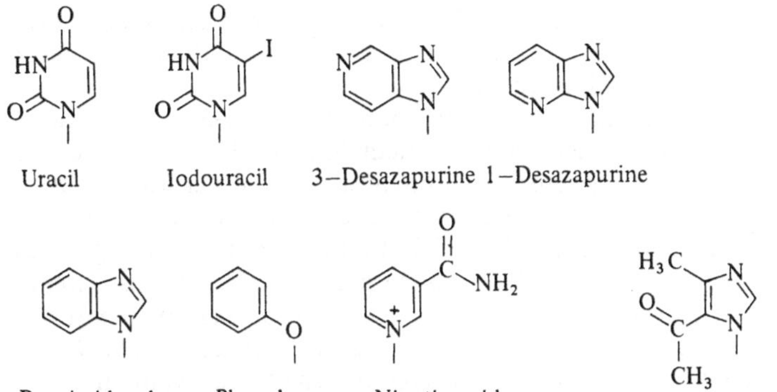

Uracil   Iodouracil   3–Desazapurine   1–Desazapurine

Benzimidazole   Phenol   Nicotinamide   5–Acetyl–4–methylimidazole

Fig. 8

In coenzymes that contain these ring systems the intramolecular interaction is weaker than in the above-mentioned coenzyme analogues that contain other purine structures. Energy transfer was observed in the case of dihydro-

Table 2. Coenzyme properties of analogues with substitution of the nonfunctional ring system

Enzymes	L-ADH		Y-ADH		LDH	
	$K_M \times 10^4$	V_{max}	$K_M \times 10^4$	V_{max}	$K_M \times 10^4$	V_{max}
Nicotinamide-uracil-dinucleotide	3.7	230	50	2500	5.6	15000
Nicotinamide-5-iodouracil-dinucleotide	3.6	280	20	3000	3.8	12000
Nicotinamide-3-desazapurine-dinucleotide	0.6	260	40	15000	6.2	20000
Nicotinamide-1-desazapurine-dinucleotide	3.3	250	30	4000	2.8	15000
Nicotinamide-benzimidazole-dinucleotide	3.7	200	38	1300	6	15000
Nicotinamide-5-acetyl-4-methylimidazole dinucleotide	0.6	260	40	16000	5	25000
Nicotinamide-phenoxy-dinucleotide	6	240	50	7000	8.7	15000

nicotinamide-benzimidazole dinucleotide and dihydronicotinamide-1-desa-zazapurine dinucleotide. The catalytic activities of these derivatives are shown in Table 2.

The coenzyme analogues exhibit different activities with different enzymes. Thus, for catalytic activity in the lactate dehydrogenase system, it is sufficient that an aromatic ring system be present in the nonfunctional part of the molecule. The Michaelis constants K_M increase, but the turnover numbers are not reduced. The coenzyme properties of nicotinamide-3-desazapurine dinucleotide and nicotinamide-5-acetyl-4-methyl-imidazole dinucleotide in the test with alcohol dehydrogenase from horse liver are shown in Table 2. In these cases the Michaelis constants are not reduced, and are comparable to those obtained in the presence of $NAD^{\oplus}$. Both these coenzyme analogues contain a polarizable group which relative to the structure of adenine is present at the N-1 position. The importance of this ring nitrogen was first reported by Kaplan and coworkers[25] in their studies on the catalytic activities of N-6-hydroxyethyl-NAD and N-1-hydroxyethyl-NAD. The N-1 derivative had much less cocatalytic activity than its isomer, the N-6-derivative.

The symmetrical coenzyme derivative bis-nicotinamide dinucleotide has little coenzyme activity[26] and in various enzyme tests is converted at a very low rate to the symmetrical bis-dihydronicotinamide dinucleotide (Fig. 9).

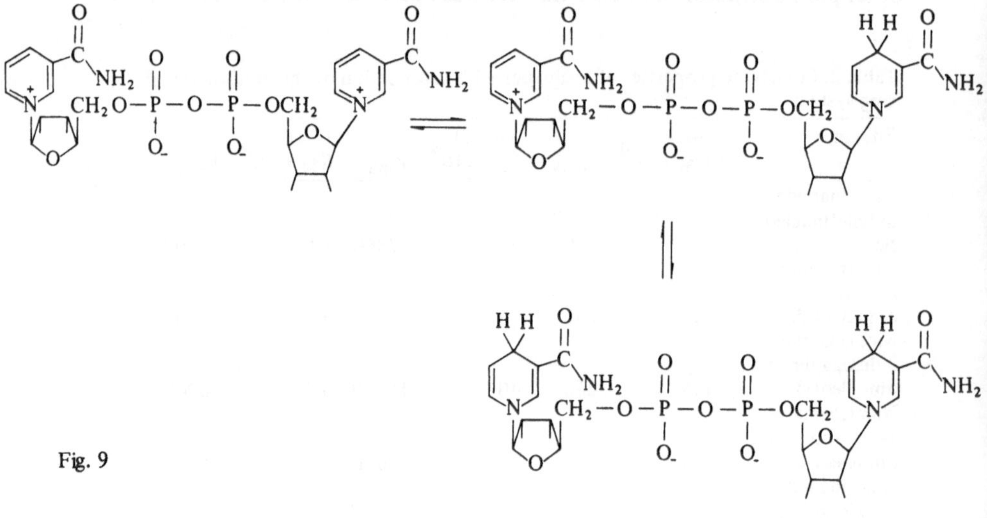

Fig. 9

The latter compound can be converted by backreaction into another nonsymmetrical form, nicotinamidedihydronicotinamide dinucleotide. The cocatalytic activity of bis-nicotinamide dinucleotide is as good as that of the coenzyme fragment containing no adenine[27]. If one of the nicotinamide rings is converted to the dihydro form, the coenzyme analogue can function as a hydrogen acceptor with an efficiency equal to that of nicotinamide-benzimidazole dinucleotide. Conversely, the semihydrogenated form, nicotinamide-dihydronicotinamide dinucleotide, is as good a hydrogen donor as dihydronicotinamide mononucleotide, as is shown in Table 3.

Table 3. Coenzyme properties as hydrogen acceptor

Enzymes	L-ADH		Y-ADH		LDH	
	$K_M \times 10^4$	V_{max}	$K_M \times 10^4$	V_{max}	$K_M \times 10^4$	V_{max}
Bis-nicotinamide dinucleotide	100	30	140	570	100	300
Nicotinamide-dihydronicotinamide-dinucleotide	4.5	200	40	1300	2	14000
Nicotinamideribose-5'-pyrophosphate-5"-ribose	40	100	100	1000	60	500
Properties as hydrogen donor						
Dihydronicotinamide-benzimidazole dinucleotide	0.4	2500	2	32000	0.3	66000
Bis-dihydronico-tinamide-dinucleotide	0.5	6300	3.5	50000	0.6	55000
Nicotinamide-di-hydronicotinamide-dinucleotide	10	120	50	900	10	3500
Dihydronicotin-amide-ribose-5'-pyrophosphate-5"-ribose	3	2300	20	10000	1	9600
Dihydronicotin-amide mononucleotide	10	30	50	40	50	1500

It is concludet from studies with various forms of the bifunctional analogues that the dihydronicotinamide part is preferentially attached at the adenine binding site, indicating its hydrophobic character. The polar pyridine ring is never bound at this position. The presence of one or both rings in dihydronicotinamide form may cause the coenzyme analogue to be bound firmly enough to allow the reaction of the other ring to proceed. To obtain the semihydrogenated form from bis-nicotinamide dinucleotide, it is necessary to bring the polar pyridine ring to the nonpolar adenine binding site. The same situation is valid for the backreaction.

The dissociation constants of nicotinamide-dihydronicotinamide dinucleotide-dehydrogenase complexes are similar to those of other coenzyme analogue complexes. It is therefore assumed that, during complex formation by the semi-reduced form, the dihydronicotinamide ring of nicotinamide-dihydronicotinamide dinucleotide readily attaches to the adenine binding site[27]. The fluorescence spectrum of the nicotinamide-dihydronicotinamide dinucleotide complex shows an energy-transfer band at 280 nm but, according to our experimental results, the dihydronicotinamide part is not fixed to the nicotinamide binding site. Perhaps there is some interaction between the tryptophan residue of the active center and the nonfunctional part of the coenzyme. Studies on the fluorescence of various dihydro coenzyme analogues and their enzyme complexes indicate that it is the nonfunctional part that influences the fluorescence of binary complexes[28]. The fluorescence of a complex is suppressed if the coenzyme analogue is not bound sufficiently at the active site, as in the case of dihydronicotinamide mononucleotide and dihydronicotinamide-ribose-5′-pyrophosphate-5″-ribose. Here, even at high concentrations of the coenzyme fragments, no energy transfer band is observed and no increase can be detected in the coenzyme fluorescence in the presence of enzyme. However, we observed in all cases a quenching of the protein fluorescence at 340 nm after irradiation at 280 nm. In the case of dihydronicotinamide-benzimidazole dinucleotide, a threefold increase in coenzyme fluorescence is seen after complex formation with the alcohol dehydrogenase of liver[28]. The energy-transfer band and the intensity of quenching of the protein fluorescence does not differ from that of the NADH-alcohol dehydrogenase complex. In other enzyme—coenzyme analogue complexes one observes a decrease in the intensity of the coenzyme complex fluorescence and the energy-transfer band.

The binding of the nonfunctional part of the coenzyme can be studied by measuring the absorption spectra of the complex against those of the isolated components. The formation of enzyme—coenzyme complexes changes the intramolecular interactions; the intensity of these interactions can be measured by comparing the spectra of dihydro coenzyme analogues and the mononucleotides. Against NADH the spectrum of an equimolar mixture of dihydronicotinamide mononucleotide and adenosine monophosphate shows rises at 335 nm and 260 nm and a minimum at 280 nm. The peaks at 320 nm and

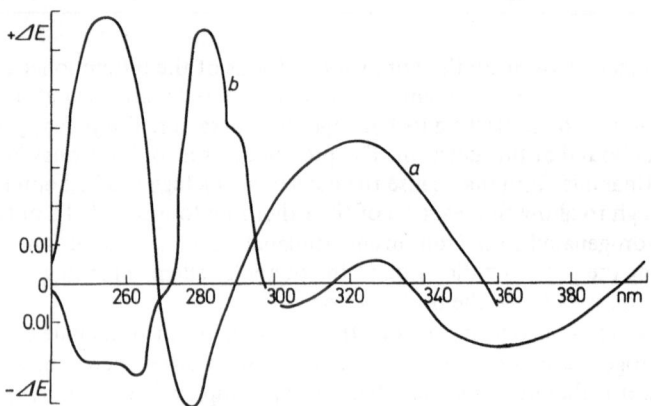

Fig. 10. Difference spectra of (a) NADH, $c = 1 \times 10^{-4}$ M against the two coenzyme fragments: adenosine monophosphate, $c = 1 \times 10^{-4}$ M and dihydronicotinamide mononucleotide, $c = 1 \times 10^{-4}$ M, pH 8.3, (b) NADH and lactate dehydrogenase against the NADH-lactate dehydrogenase complex, all components 1×10^{-4} M

260 nm are due to the reduced hypochromic effect, a result of splitting of the pyrophosphate bridge. In addition, there is a blue shift of 5 nm in the absorption band of the dihydronicotinamide part. The intramolecular interaction due to the dipole-dipole relation with the adjacent bases reduces and broadens the absorption bands of the dihydronicotinamide and adenine. If the coenzyme were bound to the active center of dehydrogenases in an open form, one would not expect an intramolecular interaction between the rings. Thus, one would expect to see changes in the absorption behavior of the difference spectra of complexed NADH and nonbound NADH (Fig. 10).

The minimum at 340 to 360 nm is thought to be due changes in the absorption of dihydronicotinamide. If one compares the spectra of complexed NADH and nonbound NADH, one can interpret the absorption changes as follows. The relationship of the adenine ring to the protein part now changes and becomes stronger than that to the dihydronicotinamide ring; hence the strong reduction of the adenine absorption band in the complex at 260 nm. A small peak in the region of 330 nm could be due to changes in the hypochromic effect of the dihydronicotinamide part, since after binding to the enzyme, absorption is shifted to shorter wavelengths[10, 29]. In the difference spectra one observes a remarkable minimum. The long-wave flank of the adenine absorption at 280 nm is very sensitive toward the nature of the solvent. The changes in this region of the spectrum can be explained on the basis of n-π* interactions, which also take place when the adenine ring is protonated in the N-1 position[30]. On the other hand, when dihydronicotinamide-benzimidazol dinucleotide is the complexing partner, new absorption bands, which

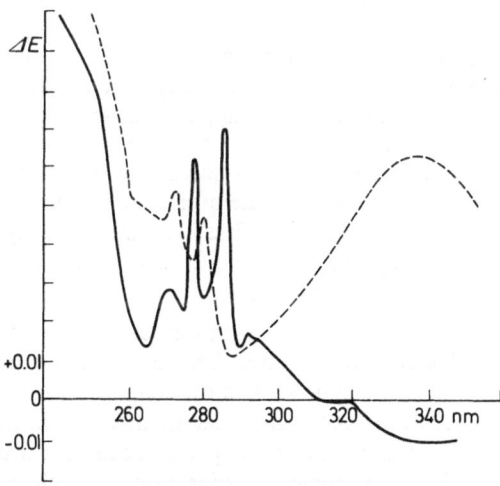

Fig. 11. Dotted line: spectrum of dihydronicotinamide-benzimidazole dinucleotide, $c = 1 \times 10^{-5}$ M, pH 8.5.
Solid line: difference spectrum of dihydronicotinamide-benzimidazol dinucleotide and lactate dehydrogenase against the complex[28]

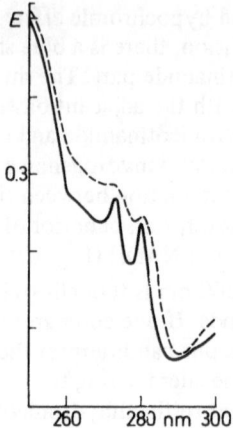

Fig. 12. Solid line: spectrum of benzimidazolriboside in water; dotted line: in dioxane—water mixture (v : v = 50 : 50), $c = 1 \times 10^{-4}$ M

coincide with the absorption maxima of benzimidazole, occur in the region 240 to 280 nm (Figs. 11 and 12). Protonation of the benzimidazol ring is not possible; the changes in the absorption bands are due to the disappearance of the fine structure of benzimidazole absorption after binding to the active center. A change in the absorption of benzimidazole in a nonpolar solvent is similarly an indication of the hydrophobic character of the adenine binding site.

We thought the spectral changes might be induced by aromatic amino-acid residues; we therefore used the coenzyme fragment dihydronicotinamide-ribose-5'-pyrophosphate-5"-ribose and the analogue bis-dihydronicotinamide dinucleotide with lactate dehydrogenase, but these neither increased the extinction at 280 nm nor decreased the extinction at 260 nm. Complexes of these analogues with alcohol dehydrogenase of liver, and to some extent of yeast, did increase extinction at 280 nm. Thus, with alcohol dehydrogenase, but not with lactate dehydrogenase, complex formation does involve a change in the conformation of aromatic amino acids.

Complexes of lactate dehydrogenase with various coenzyme analogues exhibit differences in degree of dissociation. The role of amino-acid residues which have dissociable groups can be studied by examining the stability of complexes at various pH values. Analogues that are modified in the nonfunctional part all exhibit a similar type of pH dependence in their dissociation constants. Using NADH or dihydronicotinamide-benzimidazole dinucleotide as complexing partner, we found that with alcohol dehydrogenase from yeast there is a significant decrease in the stability of the complexes between pH 6 and 7. This indicates that the binding of the coenzyme involves a side chain with a pK of 6.5.

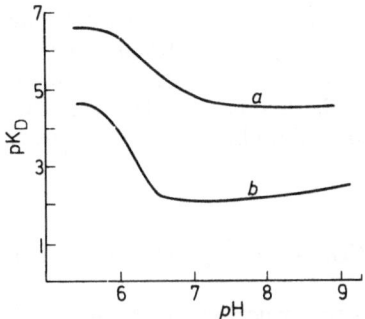

Fig. 13. pH dependence of the pK$_D$ values of complexes of alcohol dehydrogenase from yeast with (a) NADH, (b) dihydronicotinamide-benzimidazol dinucleotide

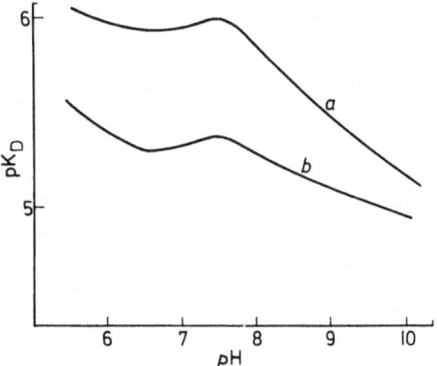

Fig. 14. pH dependence of the pK$_D$ values of complexes of alcohol dehydrogenase from liver with (a) NADH, (b) dihydronicotinamide-benzimidazole dinucleotide

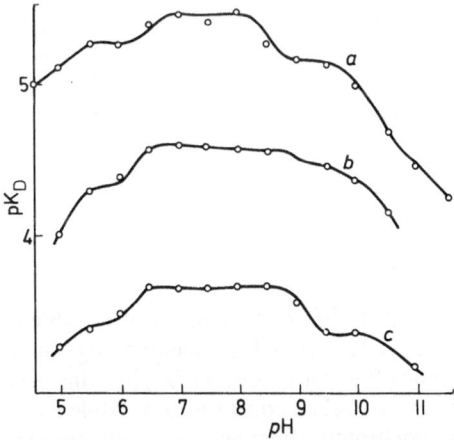

Fig. 15. pH dependence of the pK$_D$ values of lactate dehydrogenase complexes with (a) NADH, (b) dihydronicotinamide-benzimidazole dinucleotide, (c) dihydronicotinamide-ribose-5'-pyrophosphate-5''-ribose

Since the same pH dependence was observed with the benzimidazole analogue, this side chain is probably not involved in the binding of the nonfunctional part. In alkaline medium the stability of the complexes of both analogues is the same. Under these conditions the coenzyme may be held by the ionic interaction of the pyrophosphate residue to a specific group in the active center[31].

4. The Role of the Ribose Moiety in the Coenzyme Molecule

The first coenzyme analogue, which had the ribose in the nonfunctional adenosine moiety substituted by 2-deoxyribose, was synthesized by Honjo[32]. This derivative showed lower coenzyme activity than $NAD^\oplus$, so it was supposed that the coenzyme binding involves interaction of the amino-acid residues with the hydroxyl groups of the sugar. The substitution of D-ribose of adenosine by L-ribose greatly inhibits coenzyme activity[33]. The K_M and V_{max} values in this case are similar to those of the adenine-free coenzyme fragment. Complexes of the L-ribose analogue with dehydrogenases dissociate strongly. The redox potential and the intramolecular interaction between adenine and the functional ring are similar to those obtained with the natural coenzyme. Perhaps therefore the reduced coenzyme activity is due to difficulties in binding to the active center. X-ray structure analyses of binary complexes of coenzymes and dehydrogenases by Rossmann[31] have shown that, in the case of lactate dehydrogenase, the aspartic and glutamic acid residues interact with the hydroxyl groups of the sugar, so giving rise to hydrophilic regions in the active center[31]. The hydrophilic character of the binding site is also favored by the fact that there is some interaction between the coenzyme analogues nicotinamide-ribofuranosyl-[-ω-(adenine-9-yl)-n-alkyl] pyrophosphates with dehydrogenases[34].

Fig. 16 $n=1,2,3,4$

By using aliphatic chains of different lengths, it is possible to vary the distance between the pyrophosphate part and the adenine ring. Such analogues exhibit a smaller hypochromic effect than does $NAD^\oplus$, but this nevertheless indicates that the analogues have a stacked conformation. While there is no strict correlation between the hypochromic effect and the chain length of the aliphatic groups, the hypochromic effects move in a similar way to the melting points of aliphatic hydrocarbons of different chain lengths:

derivatives with butyl and ethyl chains exhibit a higher hypochromic effect than the propyl and pentyl derivatives. The fluorescence excitation spectrum complexes between the oxidized or reduced forms of these analogues with dehydrogenases resemble those of NAD$^\oplus$ and NADH enzyme complexes. The stability of the complexes decreases as the chain length of the aliphatic hydrocarbon residues increases[34].

The ethyl and butyl derivatives form comparatively stable binary complexes. The cocatalytic activity of these derivatives differs in different dehydrogenase systems (Table 4). Their reactivity in oxidized or reduced form with lactate,

Table 4. Coenzyme properties of analogues with different aliphatic hydrocarbon chains

Enzymes Length of the chain		L-ADH $K_M \times 10^3$	V_{max}	Y-ADH $K_M \times 10^3$	V_{max}	LDH $K_M \times 10^3$	V_{max}
Ethyl	($n = 1$)	3	40	20	300	5	30
Propyl	($n = 2$)	10	350	50	2000	10	150
Butyl	($n = 3$)	10	700	30	8000	6	200
Pentyl	($n = 4$)	10	3000	50	8000	10	100
Dihydroforms							
Ethyl	($n = 1$)	3	3000	1	2000	3	4000
Propyl	($n = 2$)	2	5000	4	2000	2	5000
Butyl	($n = 3$)	1	16000	3	4000	1	12000
Pentyl	($n = 4$)	1	11000	2	4000	2	12000
NADH		0.008	4000	0.05	150000	0.01	140000

malate, and alcohol dehydrogenase of yeast is very slight compared with that of NAD$^\oplus$ or NADH. In the alcohol dehydrogenase system of liver we observed high K_Ms, but the V_{max}s were extremely high except in the case of the ethyl derivative.

The substitution of ribose by other sugars on the functional nicotinamide site causes an abrupt change in the redox potential[35], which in turn influences cocatalytic behavior. The substitution of ribose by glucose in the functional part of the molecule causes a strong withdrawal of electrons from the pyridine ring, and this gives a more positive redox potential than with NAD$^\oplus$. This derivative does not exhibit any coenzyme activity[36]. The substitution of ribose by

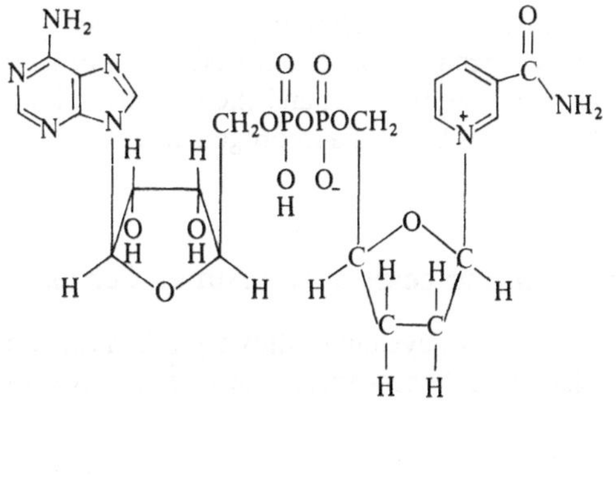

Fig. 17

a pentamethylene residue gives rise to a derivative with a more negative redox potential than $NAD^{\oplus}$[37]. Dihydronicotinamide-pentamethylene-adenosine pyrophosphate is weakly active as a hydrogen donor in enzymatic tests. The nicotinamide-2',3'-didesoxyribofuranosyl-adenosine pyrophosphate and its dihydro form have some cocatalytic activity, but it is less than that of $NAD^{\oplus}$ or $NADH$[38].

It seems, therefore, that the hydroxyl groups of ribose not only contribute to the fixation of the nicotinamide moiety to the active center, but also play an important role in the interaction of nicotinamide with the protein residues leading to its activation. The acetylation of the hydroxyl groups of the ribose of nicotinamide riboside shifts its redox potential toward the positive side[39]. The acetylpyridinium-adenine dinucleotide has a different structure from $NAD^{\oplus}$. The acid amide group in position 3 of the pyridine ring has been replaced by a methyl keto residue. This derivative has a redox potential of -257 mV against -317 mV of NAD. The substitution of ribose by an aliphatic group shifts the potential of the acetylpyridinio-alkyl analogues toward the negative side. One can therefore expect this derivative to have coenzyme activity in oxidized as well as reduced forms. The redox potential of 3-acetyl-pyridinio-n-alkyl phosphate was found to be -320 mV.

The distance between the functional acetylpyridine ring and the pyrophosphate moiety can be varied by introducing aliphatic groups of different chain length (Fig. 18). The ethyl derivative and its dihydro form do not

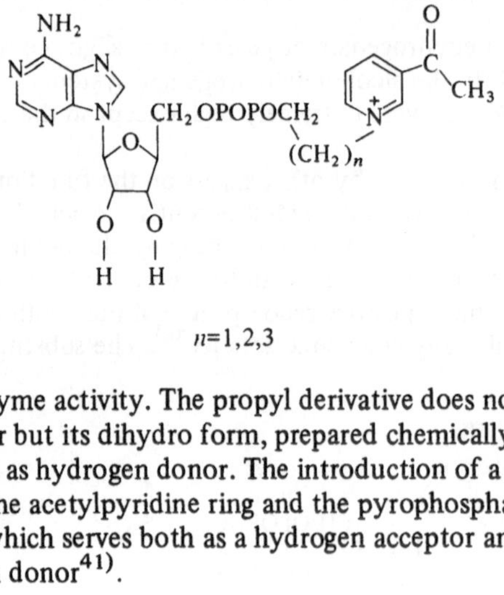

Fig. 18 $n=1,2,3$

exhibit any coenzyme activity. The propyl derivative does not function as hydrogen acceptor but its dihydro form, prepared chemically by hyposulfite treatment[40], acts as hydrogen donor. The introduction of a tetramethylene residue between the acetylpyridine ring and the pyrophosphate moiety gives a derivative that which serves both as a hydrogen acceptor and, in its dihydro form, as hydrogen donor[41].

5. Modification of Amino-acids at the Active Center

The chemical modifications of cysteine sulfhydryl, histidine imidazole and arginine residues in lactate dehydrogenase lead to a total loss of enzymatic activity[42−44, 50−53]. The catalytic activities of alcohol dehydrogenase and glyceraldehyde-3-phosphate dehydrogenase depend on cysteine sulfhydryl groups[45, 45a, 46, 54]. In the case of glutamate dehydrogenase, lysine[47] is required for catalytic activity, whereas malate dehydrogenase requires methionine[48] and histidine imidazole[49].

The treatment of methyl ketones with bromine or chlorine in acidic aqueous medium in the presence of visible light leads to the formation of halogen methyl ketones (Fig. 19). Since some of our coenzyme analogues had

methyl ketone groups, this reaction was used to convert the analogues to bromoacetylpyridinio derivatives. The monophosphates of acetylpyridinio derivatives were found to react fast and quantitatively without any degradation to give the bromoacetyl compounds. The 3-bromoacetylpyridinioadenine dinucleotide could not prepared by direct bromination of the coenzyme analogue. From the reaction mixture we isolated 3-bromoacetylpyridine and adenosine diphosphate ribose. Both derivatives are strong inhibitors of dehydrogenases[43, 55]. The synthesis of 3-chloroacetylpyridinio-adenine dinucleotide an inactivator of dehydrogenases has been described by Biellmann[56]. Westheimer[57] has described the synthesis of 3-diazoacetoxymethylpyridinio-adenine dinucleotide which, on irradiation with visible light inactivates dehydrogenases. This is due to the formation of a covalent bond between the derivative and some part of the enzyme. The 3-diazopyridinio-adenine dinucleotide has been reported by Anderson[58] and Hixson[59] to be an inactivator of dehydrogenases.

The coenzyme analogue [3-(3-acetylpyridinio)-propyl]-adenosine pyrophosphate and nicotinamide-5-acetyl-4-methylimidazole dinucleotide were
converted to their bromoacetyl derivatives, which are known to inactivate a
number of dehydrogenases[60−62]. Nicotinamide-5-bromoacetyl-4-methyl-
imidazole dinucleotide inactivates alcohol dehydrogenase of liver. If the inactivation is carried out in presence of ethanol, this analogue is incorporated into
the enzyme protein in its dihydro form[41]. Similarly, [4-(3-bromoacetylpyri-
dinio)-butyl]-adenosine pyrophosphate could be incorporated in the enzyme
protein in its dihydro form in presence of the substrate. The inactivating
properties of various coenzyme analogues are shown in Table. 5.

Table 5. Inactivation behavior of various analogues

Enzyme	L-ADH	Y-ADH	MDH_c	MDH_m	LDH	GAPDH
Nicotinamide-[5-bromo-acetyl-4-methylimidazol] dinucleotide	+	+	+	−	−	+
[3-(3-Bromoacetyl-pyridinio)-propyl]-adeno-sine pyrophosphate	−	+	−	−	−	+
[3-(4-Bromoacetylpyri-dinio)-propyl]-adenosine pyrophosphate	+	+	−	−	+	+
[4-(3-Bromoacetylpyridi-nio)-butyl]-adenosine pyrophosphate	+	+	−	−	−	+

The enzymatic activity of glyceraldehyde-3-phosphate dehydrogenase and
alcohol dehydrogenase from yeast is inhibited in the presence of most of these
coenzyme analogues. The mechanism of inactivation was studied in detail.
After incorporation of [3-(3-bromoacetylpyridinio)-propyl] adenosine pyrophosphate in dehydrogenases, the covalently bound coenzyme can be converted to its dihydro form by treatment with hyposulfite[63]. Studies on the
fluorescence spectra of the coenzyme—enzyme derivatives showed a decrease
in protein fluorescence at 340 nm after irradiation at 280 nm. The excitation
spectrum has an energy-transfer peak at 280 nm. The optical properties of the
coenzyme—enzyme derivatives are similar to those expected for the reversible
binary complexes. The complexes of nonbrominated coenzyme analogues and
dehydrogenases exhibit high dissociation constants, and the optical properties
of these complexes are not as distinct as those of the inactivated enzymes and
the natural binary complexes.

Glyceraldehyde-3-phosphate dehydrogenase is completely inactivated by
[3-(3-bromoacetylpyridinio)-propyl]-adenosine pyrophosphate. The dihydro
form, obtained by treatment with hyposulfite, does not increase the coenzyme
fluorescence, nor does formation of the NADH-glyceraldehyde-3-phosphate
dehydrogenase complex. If we assume that the inactivator first forms a reversible complex with the enzyme and then by formation of a covalent bond gives

rise to an enzyme—coenzyme derivative, we can expect a relation analogous to the Michaelis-Menten equation[64]:

$$E + I \underset{}{\overset{K_1}{\rightleftharpoons}} [E \ldots I] \xrightarrow{k_2} E\text{-}\text{-}I.$$

The inactivation constants K_I are of the same magnitude as the dissociation constants of the binary enzyme complexes measured with the nonbrominated analogue. The fact that the inactivator binds at the coenzyme binding site of the enzyme is shown by the slow reaction rate in presence of NAD$^{\oplus}$ or NADH; furthermore, the optical properties of the inactivated enzymes are similar to those of the reversible complexes. The covalent binding of the coenzyme analogue to the enzyme resembles the binding of the natural coenzyme. Thus it seems that covalent binding does not change the conformation of protein. Where dissociable protein residues are involved in the inactivation reaction information on the pK values of the relevant residues can be obtained by studying the pH dependence of the inactivation kinetics. The method is limited, however, because at pH 8 and above the inactivators are quickly hydrolyzed, whereas the dehydrogenases show some denaturation at pH 5 and below. We found that up to pH 5.7 [3-(3-bromoacetylpyridinio)-propyl]-adenosine pyrophosphate inactivates the alcohol dehydrogenase from yeast slowly but at a constant rate; at higher pH values the speed of inactivation increases. Since the turning point in the curve lies at pH 6.4, the inactivation must involve a group with this pK value. With the isomer [3-(4-bromoacetylpyridinio)-propyl]-adenosine pyrophosphate, one finds no change in the speed of inactivation between pH 5 and pH 8; however, an increase is seen at alkaline pH with nicotinamide-5-bromoacetyl-4-methylimidazole dinucleotide. The speed of inactivation of glyceraldehyde-3-phosphate dehydrogenase with [3-(3-bromoacetylpyridinio)-propyl]-adenosine pyrophosphate does increase in both a weakly acidic and an alkaline medium.

In order to study the incorporation of these derivatives into dehydrogenases, we have labeled [3-(3-bromoacetylpyridinio)-propyl]-adenosine pyrophosphate and [3-(4-bromoacetylpyridinio)-propyl]-adenosine pyrophosphate with ^{14}C in the acetyl moiety[65]. The incorporation of 4.8 moles of inactivator per mole of the tetrameric enzyme, alcohol dehydrogenase from yeast, causes total loss of its enzymatic activity. In the presence of NAD$^{\oplus}$, incorporation of 0.8 moles inactivator per mole enzyme occured; however, the incorporation of such a large amount of inactivator did not cause any significant inhibition of the enzymatic activity. In the case of glyceraldehyde-3-phosphate dehydrogenase, the incorporation of two moles of inactivator per mole of tetrameric rabbit muscle enzyme caused complete loss of enzymatic activity (Fig. 20). Probably only two of the four NAD-binding sites are occupied by the inactivator, yet reactivation of the enzyme is not possible. Equilibrium dialysis with ^{14}C-labeled NAD$^{\oplus}$ shows that the enzyme still binds two moles of NAD$^{\oplus}$. The inactive enzyme possesses the same properties as the original enzyme: hybrid bonds between yeast enzyme and rabbit-muscle enzyme after inactivation were not different from those with the native enzymes.

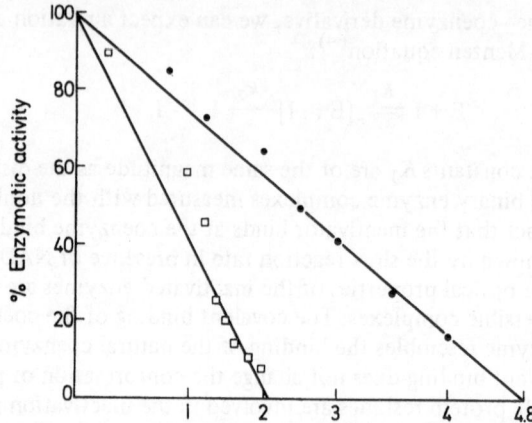

Fig. 20. Incorporation of ^{14}C-labeled [3-(3-bromoacetylpyridinio)-propyl]-adenosine pyrophosphate into alcohol dehydrogenase from yeast (●) and glyceraldehyde-3-phosphate dehydrogenase (■).[66] Abs.: μ mole inactivator/μ mole enzyme

The bond between the pyridine ring and the modified amino acid is not stable. The compounds which are prepared from 3-bromoacetylpyridine and amino acid disintegrate spontaneously into nicotinic acid[43]. Carboxymethyl derivatives of amino acids can be prepared from coenzyme—enzyme derivatives by careful oxidation with hydrogen peroxide. (Fig. 21).

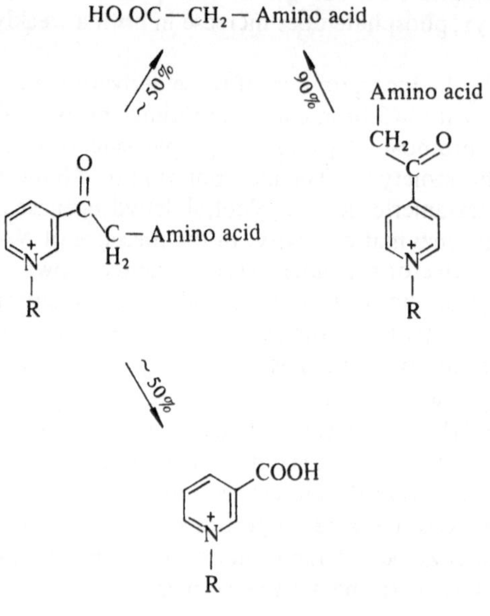

HO OC—CH$_2$— Amino acid

Fig. 21

Inactivation of alcohol dehydrogenase from yeast with ^{14}C-labeled [3-(3-bromoacetylpyridinio)-propyl]-adenosine pyrophosphate followed by oxidation showed the presence of 1-carboxymethyl histidine[66]. After inactivation of the enzyme with labeled [3-(4-bromoacetylpyridinio)-propyl]-adenosine pyrophosphate followed by oxidation, S-carboxymethyl cysteine was identified in the protein. In the case of glyceraldehyde-3-phosphate dehydrogenase, treatment with either coenzyme analogue leads to the modification of the cysteine residue. Treatment with [^{14}C]nicotinamide-5-bromo-4-methylimidazole dinucleotide did not reveal any modified amino-acid-residues. The labeled nicotinamide residue split off during the recovery of the inactivated enzyme. Attempts to synthesize an inactivator labeled with a ^{14}C-acetyl residue did not give satisfactory yields. If the enzyme–coenzyme derivative was treated with tritiated sodium boron hydride, tritium could be introduced (Fig. 22). Studies with

Fig. 22

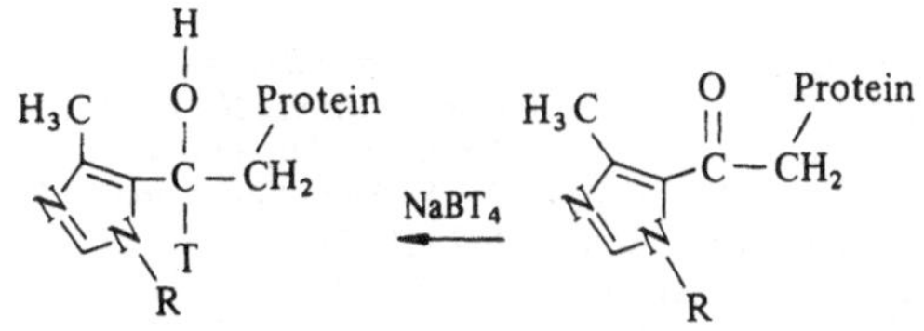

^{3}H-labeled model compounds prepared from the inactivator and N-acetylamino acids showed that only histidine and cysteine were labeled in the protein.

6. Conclusion

Synthetic coenzyme analogues have proved useful in elucidating the chemical and physical nature of the active center in dehydrogenases involved in complex formation. The various structural entities of the coenzyme molecule have different and specific binding sites that interact quite specifically with the amino acid residues of the enzyme.

Interactions between enzymes and synthetic coenzyme analogues induce specific changes in fluorescence, spectral absorption, and dissociation constants. These changes in physical properties reflect the cocatalytic activity of the coenzyme analogues. Thus, complexes between coenzyme analogues and enzymes may in some cases display greater enzymatic activity than complexes between the natural coenzyme and enzymes. In most cases, however, complexes of coenzyme analogues have less activity than normal complexes.

The insertion of reactive groups into the coenzyme analogues modifies the composition of the amino acids involved in the binding and activation of the coenzyme. The covalently bound coenzyme analogue under these conditions shows no catalytic activity, but its optical properties are similar to those of the reversible complexes of NAD$^\oplus$ or NADH with the enzyme.

7. References

[1] Harden, A., Young, W. J.: Proc. Chem. Soc. *21*, 189 (1905).
[2] Plapp, B. V.: J.biol. Chemistry *245*, 1727 (1970).
[3] Warburg, O., Christian, W., Griese, A.: Biochem. Z. *282*, 157 (1935).
[4] Siegel, J. M., Montgomery, G. A., Bock, R. M.: Arch. Biochem. Biophys. *82*, 288 (1958).
[5] Weber, G.: Nature *180*, 1409 (1957).
[6] Kaplan, N. O.: Steric course of microbiological reactions (eds. F. E. W. Wolsten-holme, C. M. O'Connor), p. 37. London: Churchill 1959.
[7] Teale, F. W. J., Weber, G.: Biochem. J. *72*, 15P (1959).
[8] Velick, S. F.: J.biol. Chemistry *233*, 1455 (1958).
[9] Scott, T. G., Spencer, R. D., Leonard, N. J., Weber, G.: J. amer. chem. Soc. *92*, 687 (1970).
[10] Gross, D. G., Fisher, H. F.: Biochemistry *8*, 1147 (1969).
[11] Anderson, B. M., Ciotti, C. J., Kaplan, N. O.: J. biol. Chemistry *234*, 1219 (1959).
[12] Burton, R. M., Kaplan, N. O.: Arch. Biochem. Biophys. *70*, 107 (1957).
[13] Wallenfels, K., Diekmann, H.: Liebigs Ann. Chem. *621*, 166 (1959).
[14] Plaut, G. W. E., Plaut, K. A.: Arch. Biochem. Biophys. *48*, 189 (1954).
[15] Moffat, J. G., Khorana, H. G.: J. amer. chem. Soc. *80*, 3756 (1958).
[16] Hughes, N. A., Kenner, G. W., Todd, A. R.: J. chem. Soc. London 3733 (1957).
[17] Ikehara, M.: Chem. and Pharm. Bull. *8*, 367 (1960).
[18] Davoll, J., Lowy, B. A.: J. amer. chem. Soc. *73*, 1650 (1951).
[19] Robins, M. J., Robins, R. K.: J. amer. chem. Soc. *87*, 4934 (1965).
[20] Tener, G. M.: J. amer. chem. Soc. *83*, 159 (1961).
[21] Seita, T., Yamanchi, K., Kinoshita, M., Imoto, M.: Bull. chem. Soc. Japan, *45*, 926 (1972).
[22] Leonard, N. J., Laursen, A. R.: Biochemistry *4*, 365 (1965).
[23] Woenckhaus, C.: Chem. Ber. *97*, 2439 (1964).
[24] Fawcett, G. P., Kaplan, N. O.: J. biol. Chemistry *237*, 1709 (1962).
[25] Windmueller, H. G., Kaplan, N. O.: J. biol. Chemistry *236*, 2716 (1961).
[26] Pfleiderer, G., Sann, E., Ortanderl, F.: Biochim. Biophys. Acta *73*, 39 (1963).
[27] Woenckhaus, C., Scherr, D.: Hoppe-Seyler's Z. Physiol. Chem. *354*, 53 (1973).
[28] Scherr, D., Jeck, R., Berghäuser, J., Woenckhaus, C.: Z. Naturforsch. *28c*, 247 (1973).
[29] Theorell, H.: Advances in Enzymology *20*, 31 (1958).
[30] Deavin, A., Fisher, R. F., Kemp, C. M., Mathias, A. P., Rabin, B. R.: Europ. J. Biochem. *7*, 21 (1968).
[31] Adams, M. J., Buehner, M., Chandrasekhar, K., Ford, G. C., Hackert, M. L., Liljas, A., Rossmann, M. G., Smiley, I. E., Alliso W. S., Everse, J., Kaplan, N. O., Taylor, S. S.: Proc. Natl. Acad. Sci. USA *70*, 1968 (1973).
[32] Honjo, M., Furukawa, Y., Moriyama, H., Tanaka, K.: Chem. and Pharm. Bull. *11*, 712 (1963).
[33] Holý, A.: personal communication.
[34] Jeck, R., Wilhelm, G.: Liebigs Ann. Chem. *1973*, 531.
[35] Karrer, P., Ringier, B., Büchi, J., Fritzsche, H., Solmsen, U.: Helv. Chim. Acta *20*, 55 (1937).
[36] Woenckhaus, C., Volz, M., Pfleiderer, G.: Z. Naturforsch. *19b* 467 (1964).
[37] Goebbeler, K. H., Woenckhaus, C.: Liebigs Ann. Chem. *700*, 180 (1966).
[38] Woenckhaus, C., Jeck, R.: Liebigs Ann. Chem. *736*, 126 (1970).
[39] Pfleiderer, G., in: 14th Coll. Gesellschaft für physiologische Chemie, Mosbach, 1963, p. 300. Berlin – Heidelberg – New York: Springer 1964.
[40] Beisenherz, G., Boltze, H. J., Bücher, T., Czok, R., Garbade, K. H., Meyer-Arendt, E., Pfleiderer, G.: Z. Naturforsch. *8b*, 555 (1953).
[41] Jeck, R., Zumpe, P., Woenckhaus, C.: Liebigs Ann. Chem. 961, 1973.

42) Fondy, T. P., Everse, J., Driscoll, G. A., Castillo, F., Stolzenbach, F. E., Kaplan, N. O.: J. biol. Chemistry 240, 4219 (1965).
43) Woenckhaus, C., Berghäuser, J., Pfleiderer, G.: Hoppe-Seyler's Z. Physiol. Chem. 350, 473 (1969).
44) Berghäuser, J., Falderbaum, I.: Hoppe-Seyler's Z. Physiol. Chem. 352, 1189 (1971).
45) Li, T. K., Vallee, B. L.: Biochem. Biophys. Res. Comm. 12, 44 (1963).
45a) Harries, J. I., Godwin, T. W., Hartley, B. S.: Structure and activity of enzymes, p. 97. New York: Academic Press 1964.
46) Boyer, P. D.: J. amer. chem. Soc. 76, 4331 (1954).
47) Holbrook, J. J., Jeckel, R.: Biochem. J. 111, 689 (1969).
48) Lescovac, V., Pfleiderer, G.: Hoppe-Seyler's Z. Physiol. Chem. 350, 484 (1969).
49) Anderton, B. H.: Europ. J. Biochem. 15, 562 (1970).
50) Holbrook, J. J.: Biochem. Z. 344, 141 (1966).
51) Berghäuser, J., Falderbaum, I., Woenckhaus, C.: Hoppe-Seyler's Z. Physiol. Chem. 352, 52 (1971).
52) Baker, B. R., Bramhall, R. R.: J. Medicinal Chemistry 15, 230 (1972).
53) Eisele, B., Wallenfels, K.: Pyridine nucleotide-dependent dehydrogenases (ed. H. Sund), p. 91. Berlin – Heidelberg – New York: Springer 1970.
54) van Eys, J., Kretszchmar, R., Tseng, N. S., Cunningham, L. W.: Biochem. Biophys. Res. Comm. 8, 243 (1962).
55) Forlano, A.: J. Pharm. Sciences 56, 763 (1967).
56) Biellmann, J. F.: personal communication, French patent: Fr-2105099-Q.
57) Brown, D. T., Hixson, S. S., Westheimer, F. H.: J. biol. Chemistry 246, 4477 (1971).
58) Fisher, T. L., Vercellotti, S. V., Anderson, B. M.: J. biol. Chemistry 248, 4293 (1973).
59) Hixson, S. S., Hixson, S. H.: Photchem. Photobiol. 18, 135 (1973).
60) Woenckhaus, C., Zoltobrocki, M., Berghäuser, J.: Hoppe-Seyler's Z. Physiol. Chem. 351, 1441 (1970).
61) Woenckhaus, C., Schättle, E., Jeck, R., Berghäuser, J.: Hoppe-Seyler's Z. Physiol. Chem. 353, 559 (1972).
62) Woenckhaus, C., Zoltobrocki, M., Berghäuser, J., Jeck, R.: Hoppe-Seyler's Z. Physiol. Chem. 354, 60 (1973).
63) Sokolovsky, M., Riordan, R. J., Vallee, B. L.: Biochem. Biophys. Res. Comm. 27, 20 (1967).
64) Gold, A. M., Fahrney, D.: Biochemistry 3, 783 (1964).
65) Woenckhaus, C., Jeck, R., Schättle, E., Dietz, G. Jentsch, G.: FEBS-Letters 34, 175 (1973).
66) Chrestfield, A. M., Stein, H. W. , Moore, S.: J. biol. Chemistry 238, 2413 (1963).

Received January 28, 1974

D.F.H. Wallach

The Plasma Membrane: Dynamic Perspectives, Genetics and Pathology

27 figures. XI, 186 pages. 1972
(Heidelberg Science Library,
Vol. 18). DM 21,–
ISBN 3-540-90047-0

Distribution rights for
U.K., Commonwealth, and the
Traditional British Market
(excluding Canada): English Universities Press Ltd., London

Thin-Layer Chromatography

A Laboratory Handbook

Editor: E. Stahl
Translator: M.R.F. Ashworth
Second edition, fully revised and
expanded
241 figures and 3 plates in color
XXIV, 1041 pages. 1969
Cloth DM 140,–;
ISBN 3-540-04736-0

Distribution rights for
U.K., Commonwealth, and the
Traditional British Market
(excluding Canada): Allen & Unwin
Ltd., London

**From the reviews of the 1st English
edition:** "The name of Stahl has
become acknowledged worldwide
for his development of the thin-
layer technique as we know it to-day.
Consequently, a handbook edited
by him must inevitably be regarded
as a classic in the field. There is so
much useful information collected
in this book that it is difficult to
discuss adequately all that has been
included . . . It should be a book
ʻon hand' in libraries and many
laboratories."
The Analyst & Analytical Abstracts.

H. Determann

Gel Chromatography Gel Filtration, Gel Permeation, Molecular Sieves.

A Laboratory Handbook

Translated by E. Gross, J.M. Harlin
Second edition
40 figures. XII, 202 pages. 1969
Cloth DM 38,–. ISBN 3-540-04450-7

Springer-Verlag Berlin Heidelberg New York

Recent Results in Cancer Research

Fortschritte der Krebsforschung/
Progrès dans les recherches
sur le cancer

Edited by numerous experts.
Editor in chief: P. Rentchnick
Sponsored by the Swiss League
against Cancer. Special subscription
price, 20 % below list price, applies
when complete set is purchased

Vol. 20: Rubidomycin
A New Agent against Cancer
Editors: J. Bernard, R. Paul, M. Boiron,
C. Jacquillat, R. Maral
68 figures (8 in color). XIV, 181 pages
1969. Cloth DM 53,—
ISBN 3-540-04682-8

**Vol. 21: Scientific Basis of Cancer
Chemotherapy.** Editor: G. Mathé
60 figures. IX, 96 pages. 1969
Cloth DM 31,—. ISBN 3-540-04683-6

**Vol. 22: P. Koldovský: Tumor Specific
Transplantation Antigen**
19 figures. V, 75 pages. 1969
Cloth DM 26,50. ISBN 3-540-04684-4

**Vol. 25: P. Roy-Burman: Analogues
of Nucleic Acid Components**
Mechanisms of Action
41 figures. XI, 111 pages. 1970
Cloth DM 31,—. ISBN 3-540-04990-8

**Vol. 28: E.S. Meek: Antitumour and
Antiviral Substances of Natural Origin**
VII, 78 pages. 1970. Cloth DM 18,—
ISBN 3-540-04993-2

**Vol. 34: Chemistry and Biological
Actions of 4-Nitroquinoline 1-Oxide**
Editors: H. Endo, T. Ono, T. Sugimura
12 figures. XII, 101 pages. 1971
Cloth DM 36,—. ISBN 3-540-05230-5

**Vol. 43: Nomenclature, Methodology
and Results of Clinical Trials in Acute
Leukemias**
Workshop held June 19 and 20, 1972 at
the Centre National de la Recherche
Scientifique (C.N.R.S.), France.
Editor: G. Mathé, P. Pouillart,
L. Schwarzenberg
79 figures (some in color). 66 tab.
IX, 168 pages. 1973. Cloth DM 58,—
ISBN 3-540-06401-X

Vol. 44: Special Topics in Carcinogenesis
Symposium of the „Gesellschaft zur
Bekämpfung der Krebskrankheiten
Nordrhein-Westfalen, e.V. "Düsseldorf,
24 - 25 March, 1972.
Editor: E. Grundmann
54 figures. VIII, 188 pages. 1974
Cloth DM 58,—. ISBN 3-540-06460-5

**Vol. 45: P. Koldovský: Carcinoembryonic
Antigens**
4 figures. VII, 70 pages. 1974
Cloth DM 38,—. ISBN 3-540-06640-3

**Vol. 48: Platinum Coordination Complexes
in Cancer Chemotherapy**
Editors: T.A. Connors, J.J. Roberts
93 figures. Approx. 220 pages. 1974
Cloth DM 68,—. ISBN 3-540-06793-0

**Vol. 49: Complications of Cancer
Chemotherapy**
Editors: G. Mathé, R.K. Oldham
33 figures. Approx. 150 pages. 1974
Cloth DM 58,—. ISBN 3-540-06804-X

Springer-Verlag
Berlin Heidelberg New York
München Johannesburg London Madrid
New Delhi Paris Rio de Janeiro Sydney
Tokyo Utrecht Wien